Build Your Own CNC Machine

Patrick Hood-Daniel
James Floyd Kelly

Apress®

Build Your Own CNC Machine

ISBN-13 (pbk): 978-1-4302-2489-1

ISBN-13 (electronic): 978-1-4302-2490-7

Printed and bound in the United States of America 9 8 7 6 5 4 3 2 1

President and Publisher: Paul Manning
Lead Editors: Jonathan Gennick
Technical Reviewers: Darrell A. Kelly, Jim Burt
Editorial Board: Clay Andres, Steve Anglin, Mark Beckner, Ewan Buckingham, Tony Campbell, Gary Cornell, Jonathan Gennick, Michelle Lowman, Matthew Moodie, Jeffrey Pepper, Frank Pohlmann, Ben Renow-Clarke, Dominic Shakeshaft, Matt Wade, Tom Welsh
Coordinating Editor: Kelly Moritz
Copy Editor: Damon Larson
Compositor: Lynn L'Heuruex
Indexer: Ann Rogers/Ron Strauss
Artist: April Milne
Cover Designer: Kurt Krames

Distributed to the book trade worldwide by Springer-Verlag New York, Inc., 233 Spring Street, 6th Floor, New York, NY 10013. Phone 1-800-SPRINGER, fax 201-348-4505, e-mail orders-ny@springer-sbm.com, or visit http://www.springeronline.com.

For information on translations, please e-mail info@apress.com, or visit http://www.apress.com.

Apress and friends of ED books may be purchased in bulk for academic, corporate, or promotional use. eBook versions and licenses are also available for most titles. For more information, reference our Special Bulk Sales–eBook Licensing web page at http://www.apress.com/info/bulksales.

Patrick wishes to dedicate this book to his father – "Everyday I attempt to approach his level of logic and perfection. Rest in peace."

Jim dedicates this book to his dad – "I've learned a lot from you and enjoyed the company."

Contents at a Glance

Contents

About the Authors

Patrick Hood-Daniel is a hobbyist. In his day-job, he is an urban designer trained in architecture at the University of Miami, School of Architecture and urban design at the University of California at Berkeley. Patrick also serves as a college level instructor for various creative technical fields of mechanical engineering, architectural design, industrial design, civil engineering, materials and manufacturing, urban design and construction management. But in his spare time, Patrick is a hobbyist who puts skills from a previous career as a computer programmer to good use in building and operating computer numerical controlled (CNC) fabrication machines. He is the creative force behind www.buildyourcnc.com and is well known for designing CNC machines that can be built at low cost by normal people, and without any special or expensive tools.

James Floyd Kelly is a freelance writer living in Atlanta, Georgia, with degrees in English and Industrial Engineering. James has written on topics including LEGO robotics, building custom computers and free software. His books include *Don't Spend a Dime: The Path to Low-Cost Computing*, *Ubuntu on a Dime*, and *LEGO MINDSTORMS NXT 2.0: The King's Treasure*, the sequel to *LEGO MINDSTORMS NXT: The Mayan Adventure*. He is also the creator and editor-in-chief of the most popular NXT robotics blog, www.thenxtstep.com, with over 45,000 readers visiting each month.

About the Technical Reviewers

Jim Burt was born and raised in Goshen, Indiana. After being discharged after serving 6 years in the Navy he was employed with Westinghouse for 29 years in Pensacola, Florida, where he worked as a machinist and tool machine mechanic. "Woodworking has been a hobby all my life and I love it." Jim is 70 years old, and has been married for 52 years. He has three sons, four grandsons, three great grand daughters and one great grandson.

Darrell A. Kelly is a retired chemist (PhD) with over 30 years of industrial experience and 14 years of college teaching experience. Along with his wife of 42 years he helped raise two sons and a daughter—a writer, a CPA and an accountant, respectively. During those years he has followed three hobbies—maintaining their four acre estate, raising a vegetable garden each year and woodworking in his shop when time permits.

Acknowledgments

This has been one of the most challenging books to write, but it's also been relatively free of major bumps and technical glitches. And this is all due in part to a variety of individuals who have offered up their skills, their knowledge, and their advice when needed.

First, we (Jim and Patrick) would like to thank the amazing team at Apress. Books don't just come into existence on their own, and these two authors are indebted to the hard work of each and every Apress team member that was involved in this project. Two big cheers go first to Dominic Shakeshaft and Jonathan Gennick who took a chance and saw the potential for this book. Another cheer goes to our editor, Kelly Moritz, who patiently worked with her two authors to make certain every word, every figure, and every drawing was as accurate as possible... and delivered on time. We didn't always make our deadlines and sometimes we missed turning in a picture or two, but Kelly was a trooper and has done an outstanding job in making our book a better one. And speaking of making the book better, we want to offer up a huge thank you to our copyeditor, Damon Larson, who took our chapters, cleaned up the grammar and spelling, offered up some great suggestions for improvements, and was a pleasure to work with. And please flip to page 2 where you'll see a list of all the Apress folks involved in getting this book on the shelves; thank you all for helping us get the book completed.

In addition to the Apress team, we were extremely lucky to have two other individuals watching over our shoulders (sometimes literally) and making sure we were as accurate as possible with our writing. Jim Burt and Darrell Kelly (Jim's Dad) were our technical reviewers, and both of them now have their own working CNC machines. They caught our errors, asked questions, suggested clarifications, and continually looked for ways to help make the reader's job easier when it comes to building his or her own machine. Thank you both for the time you put into not only building your machines but also reading over the chapters numerous times.

Patrick wishes to thank all of the buildyourcnc.com readers and the ones that sent in their comments helping the development of the site through the years. He would also like to acknowledge his wife (Ana) for putting up with the dust and keeping his feet on the ground.

Jim would like to thank Russ Revels, Kevin Burkett, and members of the Artistic Woodturners (Pensacola, FL) for their help, support, and interest with this book. Jim also wants to thank his wife (Ashley) for her patience during those numerous weekends spent building his machine and his friend and co-author, Patrick, for helping him step into the world of CNC machining.

Introduction

Your very own CNC Machine. Don't think it's an unrealistic dream. You're holding in your hands the instruction manual for building your very own, fully functional, easy-to-maintain, and (relatively) inexpensive CNC machine. But we understand your disbelief – we've been there.

Traditionally, for hobbyists, woodworkers, and anyone interested in do-it-yourself projects, the reality of owning a CNC machine has been far-fetched. Traditional CNC machines are expensive, complicated, and typically only found in large manufacturing companies that can afford them… because, well, did we mention that they are expensive? (Small hobbyist CNC machines can run anywhere from $7000 and higher; professional machines can cost millions of dollars!)

But no more. For a fraction of the cost (how does under $800 sound?) of a commercially-available CNC machine, you, too, can own one of these wonderful devices. You can cut, drill, mill, and carve to your heart's content. You can create those things that sit currently in your imagination.

But there's a catch, obviously. The catch is you'll have to build it yourself. But before you go shaking your head and muttering about how you don't have the right tools for the job or the woodworking knowledge to get it done, just take a deep breath and pause. You can do this because every step of the process is covered in this book. A website has been created with videos, full color photos, and a discussion forum where you can post questions and get answers. Others have built this machine with a minimal number of tools and a complete lack of CNC machine experience – if they can do it, you can do it. The secret is taking your time, breaking up the project into small steps (which has been done for you with small, easy to read chapters), and knowing that help is just a website away.

Build Your Own CNC Machine delivers what the title promises. From plans to listings of hardware and tools required to step-by-step photos, you'll find everything you need in these pages to build a working CNC machine. (And because the book's photos are black and white, you can find matching enlarged, full color images on the book's website at `www.buildyourcnc.com/book.aspx` to help.)

Just imagine – your very own CNC machine in your garage, basement, or workshop. It would be nice, wouldn't it? Well, stop imagining and keep reading. And welcome to the world of do-it-yourself CNC machines.

Your CNC Machine

Chances are if you've picked up this book (or purchased a copy), then you're probably somewhat familiar with the term *CNC*. But maybe not. CNC is an abbreviation for *computer numerical control*. A *CNC machine*, then, is a machine that carves out objects in three dimensions from a solid block of material. CNC machines are commonly used in industry to produce small parts such as bicycle stems and tools. Low-cost CNC machines are increasingly used by serious hobbyists, especially woodworkers, to carve creations out of materials such as wood and aluminum.

Read through this chapter and then flip through the rest of the book—you'll start seeing pictures of an unusual device being built (beginning with Chapter 7 and ending with Chapter 17). That device is the do-it-yourself (DIY) CNC machine. The DIY part is because you're going to learn how to build one in this book. When you're done, you'll be the proud owner of a three-dimensional carving machine that can fabricate parts and objects from soft materials such as wood, plastics, and even aluminum.

What is CNC?

Computer numerical control is a very broad term that encompasses a variety of types of machines— all with different sizes, shapes, and functions. But the easiest way to think about CNC is to simply understand that it's all about using a computer as a means to control a machine that carves useful objects from solid blocks of material. For example, a CNC machine might begin with a solid block of aluminum, and then carve away just the right material to leave you with a bicycle brake handle.

CNC machines can be divided into two groups: turning machines and milling machines. A turning machine is generally made up of a device that spins a workpiece at high speed and a tool (sharp edge) that shaves off the undesired material from the workpiece (where the tool is moved back and forth and in and out until the desired form is achieved). A milling machine is a machine that has a spindle (a device similar to a router) with a special tool that spins and cuts in various directions and moves in three different directions along the x, y, and z axes.

Historically, you wouldn't actually need a computer to create forms with a turning machine or a milling machine. Adding a computer to the mix allows you to design a product on a computer first and then specify how the machine should cut this product. To design the product is to produce a computer-aided design (CAD) file. Then you specify how the machine should cut the product, and the result of that step is a computer-aided manufacturing (CAM) file (or G-Code file, or .NC file—there are many names for this type of file). This CAM file remembers all of the operations that the milling machine must follow to cut out the parts for the product. The computer tells the CNC machine how to build the part by interpreting the CAM file into signals that the CNC machine can understand.

Industrial Uses

Industrial applications for CNC machines have been chiefly based around the removal of metal to create a desired form. Metal is widely used for producing almost everything we see around us, even though these things may not be made of metal themselves. Some of the most obvious products that are made of metal are cars. The engine block and the parts within the transmission are directly produced from a CNC machine because tight tolerances are necessary (a tolerance is a range in dimensioning to which the machine must adhere). However, most of the parts of a car are not made by a CNC machine, but they have a latent connection to one. For example, how do you make a quarter panel? There is a hydraulic press with a thing called a die to create an impression in a sheet of metal. Most of the parts of the hydraulic press were made from a CNC machine. The die, the part that carries the negative form of the quarter panel and that can be replaced when design changes, was also made by a CNC machine, and then tempered for hardening and heat resistance. Even the plastic parts of a car have some connection to a CNC machine. Many of these parts were made from a mold that was created using a CNC machine.

Because CNC machines have very high precision and they can provide information back to the computer, they are also used in dimensional testing. If a switch (probe) is fastened to the location of the tool, it can analyze the measurements of a part that was produced. The machine runs this probe all over the part to confirm its desired form and measurements.

For more information on industrial uses of CNC machines, visit `www.cncinformation.com`.

Personal Uses

There is a large following by various hobbyists and DIYers around the globe interested in the concept of CNC machines. Roboticists, craftsmen, handymen, home machinists, small business owners, tech enthusiasts, backyard scientists, and artists have all discovered how a CNC machine can open doors to new designs and more detailed creations. A roboticist, for instance, will use a CNC machine to create the structural components of the robot with very high precision. Making these components by hand would be tedious and very time consuming. Using a CNC machine, the parts come out beautifully and fit together with great precision.

For the typical handyman, a great example of using a CNC machine might be designing and making cabinets for around the house. Typically, cabinets share many of the same dimensions and can be cut by a CNC machine over and over. Imagine cutting all of the drawers and cabinet lids by hand! The parts are numerous and the work would be quite tedious. But with a CNC machine, the individual pieces are cut and the cabinets assembled; no driving around looking for the right cabinets, having to special order them, and then waiting for delivery from the home improvement store. (The cabinets will need assembly, too, but with your own CNC machine, you'll find that the high cost of buying them in the store can be eliminated.)

CNC machines for personal use can be purchased from a variety of manufacturers, but many DIYers suffer from sticker shock the first time they begin shopping for a CNC machine. Prices of $3,000 and higher are typical for small, desktop versions that often come with a 12"×18" workspace, meaning you'll be limited to working on materials that fit in that small space. CNC machines with workspaces that allow for materials of 2'×4', for example, start around $7,000, and prices go much higher for larger workspace tables.

For most DIYers, owning their own CNC machine is still out of reach financially. But no longer—this book brings CNC within easy reach. If you can afford to spend $700 to $800, then you can afford to build your very own CNC machine.

Your DIY CNC Machine

With your DIY CNC machine, you're going to be able to do some amazing things—cut, drill, etch, and sculpt—with a variety of materials. In fact, author Patrick Hood-Daniel uses his own CNC machines to make more CNC machines! He has a machine cut and drill the MDF (medium-density fiberboard) parts used to build more CNC machines. (You can do this, too, but first you'll need to build your own DIY CNC machine—it all starts there.)

Your DIY CNC machine is made of MDF, a rigid material that holds up well to cutting and drilling, as well as being extremely strong and dimensionally stable (it doesn't shrink or expand with fluctuations in the weather or humidity). The MDF parts you'll be cutting and drilling are bolted together using a variety of sizes of bolts, nuts, washers, and other hardware. Finally, you'll be adding a mix of electronics and one computer to bring your DIY CNC machine to life and amaze your friends and family (who will, unfortunately, come up with all kinds of requests for you and your machine).

The DIY CNC machine isn't something with vague dimensions and a random mixture of hardware. We'll tell you exactly what to buy. You'll be cutting and drilling material from plans created by author Patrick Hood-Daniel and tested and used to build three machines; one by James Floyd Kelly, one by Darrell Kelly, and one by Jim Burt (not to mention the number of machines built by Patrick himself).

When you're done, however, you're not really done. CNC is a growing and changing technology, so the limits of what you can do with your machine are really up to you. While this book will give you the basic information to build and use your machine, you'll want to continue to improve your skills by delving deeper into the software and pushing the limits of your machine. (We'll provide you with some good resources for further research and learning later in the book.)

If you're like us, you're ready to begin. But trust us when we say that one of the best things you can do before starting to build your own CNC machine is this: read the entire book through at least once. Doing so will give you a glimpse of the final machine and a better understanding of how you'll get there. You may find, as we did, that half the fun of owning your own DIY CNC machine comes from building it.

HISTORY OF THE DIY CNC MACHINE, FROM PATRICK HOOD-DANIEL

My desire to hop on the bandwagon of this great hobby started as a means to an end. The end has not been realized because I became more interested in the CNC machine itself and want to provide simpler designs and instruction to others who wouldn't otherwise have the means to own a traditional CNC machine.

The DIY CNC community has been around for a long time; pretty much ever since the boom of the Internet. I learned most of what I know from the information on the Internet. With my prior design training, I spent quite a bit of time improving what others had created.

Through my effort to create an initial CNC machine from resources on the Internet, I found that the materials did not hold up well with use and tended to exhibit undesirable flexing. I learned through trying and experimenting . . . and discovered many things that worked and didn't work. I quickly learned, for example, to stick with MDF as the material of choice for making my CNC machines.

Over the years, I made hundreds of trips to the home improvement store (my laboratory of ideas). The components that I used to start my CNC journey included round metal bar stock and a bunch of very cheap MDF. I thought that the metal stock would have some pretty good rigidity—I mean . . . it's metal! But I was very wrong. After putting an assembly together and using the bar stock as the rail, I noticed quite a bit of flexing in the assembly. This was not going to work, so I came up with a better way. (I was deathly afraid of trying something that was not illustrated on the Internet in fear that if it wasn't done before, it wouldn't work. But I did it anyway.) I used aluminum angles as the rails and MDF as the midsection between the rails to provide the necessary rigidity. Initially, I tried the bar stock with this technique, but the bars would still flex. The aluminum rails wrapping the MDF worked perfectly and the machine was rigid and stable—perfect! Well, perfect is a subjective word here, but it was good enough for me. And I think by the time you're done following this book's instructions and building your own machine, you'll agree.

Everything from that point on became intuitive. The mechanics and motion of the machine were all designed so that the parts could be cut, drilled, and assembled using nothing more than a few simple hand tools. (I'm not kidding—the early machines were cut and drilled with nothing more than a mitre box, a small saw, and a battery-powered drill.)

This book documents my design; you'll be able to skip the frustration that I faced because this is the design I developed that worked. The DIY CNC machine fulfills my desire to provide others with a simple, elegant, and fully functional CNC machine.

What's Next?

Chapter 2 is going to give you a quick introduction to the tools and equipment you'll be needing as you build the DIY CNC machine, as well as vendors for purchasing the electronics for the machine. We highly recommend, however, that you read through the entire book before purchasing any tools or equipment; if you understand what will be accomplished in each chapter, you may be able to postpone some purchases until later in the book.

Hardware and Tools

You probably won't believe this, but the CNC machine described in this book can be built with a total of four tools (mitre box, mitre saw, tap, and electric drill/screwdriver) and a few miscellaneous items such as drill bits, tape measure, and the bolts, nuts, washers, and other small items. What this means for you is that you can realistically build your own CNC machine with a minimal amount of tool purchases. It may take a little longer to build using just the tools we've listed than if you have a fancy workshop at your disposal, but build it you can—visit www.buildyourcnc.com and you can watch over 30 videos showing you a CNC machine built using this small collection of tools.

Now, while building your own CNC machine can be done with this small tool list, that's not to say your job won't be a whole lot easier if you have access to a few more tools. So, in this chapter we're going to give you a brief rundown of some of the other tools we used during the building of three CNC machines. We're also going to give you the web sites where you can purchase the electronics used in the construction of your machine.

NOTE Notice that three CNC machines are discussed in this book. One was built by James Floyd Kelly, one of the books coauthors. The second and third machines were built by the two technical editors, Darrell A. Kelly and Jim Burt. Building multiple machines served many purposes—verifying that the plans and instructions were accurate, having a backup machine to photograph and test with in case a machine was cut wrong or assembled improperly, and just having a backup in case a machine was damaged somehow.

The Tools

We cannot predict what tools you'll have available during the building of your machine. We can, however, tell you the tools we used. Some of these tools, especially the power tools, can easily be rented (by the day or hour) at hardware stores and home centers, while others may be slightly difficult to find. And if you have access to a tool or two not mentioned here, that could make your work even easier. Just keep in mind, however, that this machine was designed so that it could be built with a minimum number of tools—if you find yourself lacking a tool described following and cannot find it (for purchase or rent), don't let that stop you; just improvise with the tools you do have. The CNC machine built in this book is extremely forgiving when it comes to small deviations in cutting and drilling; be as accurate as you can, use what you have available, and make the best of it.

Following is a list of our tools, with a few photos for clarification:

- Table saw: This is useful for cutting long lengths of MDF accurately. Depending on your skill, you can also cut multiple MDF pieces at once, guaranteeing they match in dimensions.

- Metal band saw: This is used for cutting the aluminum angled rail and lead screws.

- Hack saw: If a band saw is not available, this is the saw to use for cutting the aluminum angled rail and lead screws.

- Mitre box: This is useful for making accurate cuts in small MDF pieces.

- Hammer: This is for hammering things, obviously.

- Cordless screwdrivers: You'll need a Phillips and a slot head.

- Regular screwdrivers: Again, you'll need a Phillips and a slot head.

- Forstner drill bits: Forstner bits (see Figure 2-1) are extremely useful for counterboring as well as drilling large, smooth holes; regular drill bits can be used to drill counterbored holes, but these work much better.

Figure 2-1. *Forstner drill bits in various sizes*

- Brad point drill bits: These drill a flat-bottomed hole and have a sharp, centered tip that creates a "dimple" that can be used to center other drill bits for later drilling.

- Twisted drill bits: These are your standard drill bits and come in a range of sizes.

- Spade drill bits: This is another common variety of drill bit that is perfectly acceptable for drilling holes.
- Transfer punches: Transfer punches (see Figure 2-2) are available in different diameters. These tools have a sharp point on the end; inserting them into existing drilled holes will allow you to make a "dimple" in a second piece of MDF, giving you an accurate point to drill on the second piece of MDF.

Figure 2-2. *Transfer punches let you mark other pieces accurately for drilling.*

- Magnetic bowl: This is a small bowl that can keep your nuts and bolts from falling all over the floor.
- 1/2" power drill: Having a drill that can handle larger-diameter drill bits will be very useful during the build.
- Drill press: Useful for drilling straight holes (vertically) through material. A drill press also provides a small table to clamp MDF and aluminum rail to when drilling.

- Wrenches: You'll need wrenches for 1/4" nuts.
- Detail metal ruler: This is a special type of ruler (see Figure 2-3) with marks that allow you to make extremely straight lines for cutting and points for drilling. Measuring and marking increments of 1/8", 1/16", 1/32", and 1/64" are possible with these rulers.

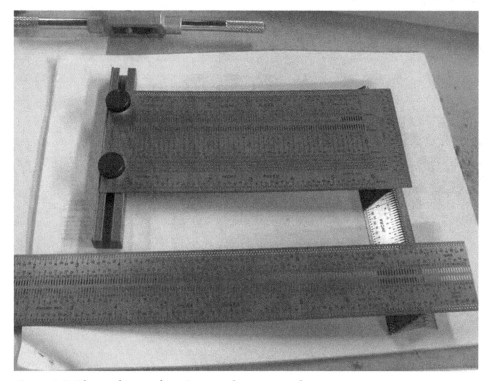

Figure 2-3. *These rulers are from Incra and are extremely accurate.*

- Tape measure: This tool is highly recommended as you'll be doing a lot of measuring in this project.
- Metal square: A 90 degree square will come in handy for ensuring that parts are joined properly.

- Tapping bit and holder: A tapping bit (see Figure 2-4) is used to cut thread into the aluminum angled rail so a bolt can be screwed in.

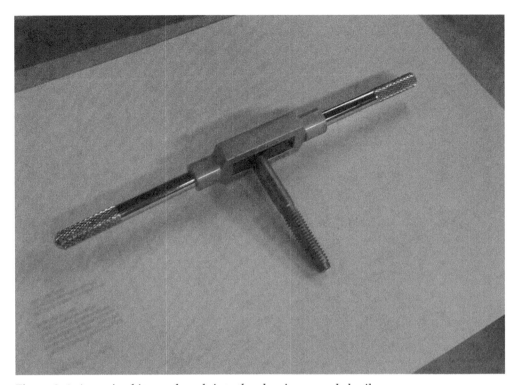

Figure 2-4. *A tapping bit cuts threads into the aluminum angled rail.*

- Center punch: This little tool allows you to make a small indentation in wood and metal to mark where to drill.
- C clamps: Clamps will come in very useful for holding parts together as you cut or drill them.
- Bar clamps (of assorted sizes): These larger clamps will come in handy later when your CNC machine begins to get bigger and you need clamps that can stretch wider and longer.
- Router: This is required for building your CNC machine. You'll need to purchase a laminate router (also called a hand router) to use with the chamfer bit.

- Chamfer bit: For this project, you'll need a 45 degree chamfer bit (see Figure 2-5); this router bit allows you to trim the edge of a piece of MDF with a 45 degree angle.

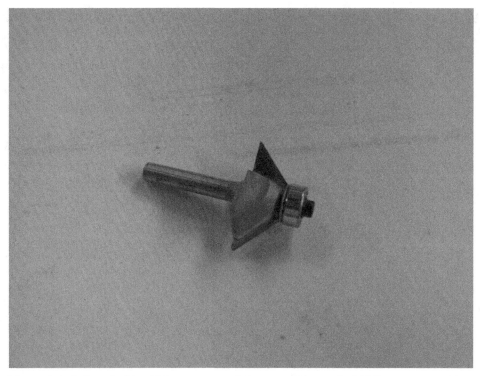

Figure 2-5. *A chamfer bit will let you cut a 45 degree surface on the MDF.*

- Countersink bits: These are used on the CNC machine's tabletop so the tapered head bolts won't stick out above the surface.

- Metal file: This is used to smooth rough metal cuts such as the rails cut from aluminum.

- Sanding block: This is used to defeather MDF and remove rough edges.

- Soldering gun: You can use wire nuts (various sizes) to connect two pieces of wire, but soldering is still the best way to make a strong, reliable connection between two pieces. The wire used in this project (18 gauge) may be too thin/delicate for a wire nut to be reliable.

- Solder: Along with a soldering gun, you'll need plenty of solder during the electronics assembly phase of the project.

- Wire cutter and stripper: Most of the wire you'll use during the electronics assembly will need to have the outer sheath removed (stripped) and cut to specific lengths.
- Third-hand clamp: This is useful for holding wire while you solder—a third-hand clamp typically has a heavy base (see Figure 2-6), at least two alligator clips, and often a magnifying glass.

Figure 2-6. *A third-hand clamp is very useful when soldering.*

- Multimeter: You'll use this tool to verify the voltage and amperage of some of the electronic devices used during the electronics assembly phase.
- This is our recommended list of tools, but it's certainly not exhaustive. There are likely tools you have (or have access to) that can make building your CNC machine that much easier. (And if you have a method or tool that works better than we describe, please let us know by sharing your information at the book's official discussion forum: www.buildyourcnc.com/book.aspx.)

The Electronics Vendors

Although it may be possible to locate other vendors for the electronics used in this CNC machine, we can provide you with three possible sources for ordering:

- Keling Technology (www.kelinginc.net): Keling is where the electronics for the three test machines were ordered. Shipping took about one week and all the parts were wrapped securely. Keling also provides downloadable documentation on most everything they sell.

- CNC4PC (www.cnc4pc.com): This is another source for some (but not all) of the electronics we specify in Chapter 6.

- www.buildyourcnc.com: Coauthor Patrick Hood-Daniel has been packaging and selling bundles of electronics (as well as hardware) for building CNC machines, and the components used in Chapter 6 can be purchased directly from his web site.

Chapter 6 provides a complete list of the electronics components you'll be needing, including model numbers. If you should decide to order slightly different components, be sure that you can obtain documentation on the parts; different components may use different colored wires, different naming conventions for labels, and so on. You may have to do some research or make some technical support calls to make certain you can match up your electronics to the electronics we describe in Chapter 6.

What's Next?

Chapters 3 through 5 are going to provide you with some information that you'll find relevant throughout the CNC machine build. There are many tasks you'll perform repeatedly, and rather than annoy you with repeated warnings and "do it this way" instructions, we're providing what we feel are some good rules of thumb to remember as you build (in Chapter 3), as well as some instructions for two tasks—joining MDF pieces and making bolt-bearing-nut assemblies—in Chapters 4 and 5.

Tips and Advice

So you've made the decision to build your own CNC machine. Congratulations. The remaining chapters in this book will provide you with the information to do just that. But before we begin, let's take some time to talk about this project because, frankly, it's a big one! The authors of this book have gone through this process (and in some instances, more than once), and there's a lot of lessons that have been learned—often from mistakes!

This chapter is going to provide you with tips, advice, and some much-needed encouragement. The completed CNC machine can be a bit overwhelming the first time you see a picture of it, but keep in mind that the CNC machine you'll be building is nothing but a large assembly consisting of smaller assemblies (and just a few of those to boot). Like the old joke "How do you eat an elephant?" you're going to see that the solution to building a CNC machine is in "small bites."

We highly encourage you to refer back to this chapter as you build your CNC machine. Those of us who have already built this machine have had moments of frustration. We've also made mistakes (and we'll tell you where in later chapters so you don't make the same ones). But we've also learned a few secrets and figured out a few helpful tips that we're happy to share with you so you, too, can have one of these great little machines to call your own.

Cut Once

In later chapters, you're going to be measuring and cutting a lot of parts. One thing we don't want to do in every chapter and in every paragraph is annoy you by repeating ourselves over and over. Only when we feel it's important that something be repeated or emphasized will we possibly assume the role of teacher and hammer in some concept again and again.

So, now that we've gotten that out of the way, we want to go ahead and give you the first bit of advice that you've probably heard numerous times in your life but applies so aptly here:

Measure twice, cut once

Your CNC machine will be made of MDF, a material that is very strong and easy to cut, drill into, and paint. But one thing you most certainly cannot do with MDF is join two pieces back together that have been cut apart incorrectly.

TIP My editor politely reminded me to add an addendum to the "measure twice, cut once" rule—don't cut when distracted. It's too easy to make mistakes when your mind is on one thing and you're doing another.

The same goes for drilling holes. Many of the holes you'll be drilling into the MDF will allow for a little inaccuracy—but not much. If the holes require that you to drill 1" from the left edge and 3/4" from the top edge, there's not much you can do if you end up drilling 3/4" from the left edge and 1" from the top edge. It's likely that your only option will be to cut a new piece of MDF and drill again.

We'll remind you in the specific chapters when certain cuts need to be accurately measured. All cuts are important, but you'll find later that some parts require a bit more precision than others. Again, don't worry—we'll alert you to parts where you need to pay special attention to measuring and cutting.

So, just to be clear, when cutting and drilling the MDF for your CNC machine, remember:

Measure twice, cut once

If you're fortunate enough to have a friend or spouse nearby, it never hurts to ask them to double-check your math. If you need to cut a 3/16" strip from a 4 5/8"-wide piece of MDF, did you mark the cut for 4 7/16" or 4 1/4"? It matters! Always check your math.

Protect Yourself

Whether you'll be using a table saw or a handsaw, or a drill press or a small portable hand drill, it pays to always be diligent when working with tools. The authors of this book assume no responsibility for your eyes, fingers, and other body parts that are exposed as you cut and drill. Always read the instructions for any power tools you use, especially the safety information.

NOTE Just as your CNC machine will be using a router to do its work, you'll be using a router to do some work on the CNC machine parts (see Chapter 7). A router is a powerful and versatile power tool, but it can also be dangerous. Read the instructions, but also consider having someone demonstrate its use at a nearby hardware store if you can. You can even make this a requirement for purchasing your router; tell the salesperson you won't purchase unless someone can go over its operation and safety.

Safety glasses are another absolute requirement (see Figure 3-1). They come in a variety of shapes and tints; buy a pair that feel comfortable since you'll be wearing them a lot. Don't concern yourself about how good they look on you, however. Some of the strongest, most protective safety glasses are the most ridiculous looking when worn. When you consider that drills can fling off material (wood, metal, or other) at super-high velocities, take no chances with your vision; buy a pair of safety glasses and get in the habit of putting them on every time you enter your work area.

Figure 3-1. *Safety glasses will protect your eyes from debris flung from drills and saws.*

Protect Your Lungs

MDF is a great material to build with—it's strong, easy to cut, doesn't splinter easily like plywood, and has a smooth surface that's great for cutting and painting. But when MDF is cut or drilled, it puts out a lot of dust. And that dust is not good for your lungs. Some MDF sheets contain minute quantities of formaldehyde! You don't want to be breathing that stuff into your lungs.

Purchase a couple of breathing masks (see Figure 3-2) from your local hardware store—they're not expensive—and wear one when you're cutting or drilling MDF. They can sometimes get hot or be a little uncomfortable, but you're not going to be wearing them for long periods of time.

TIP If possible, buy the kind of mask that comes with a small piece of bendable metal that you pinch over you nose. It will prevent your breath from exiting the top of the mask and fogging up your safety goggles. You might also consider a mask that comes with a small valve on the front, which will also prevent fogging of your safety glasses.

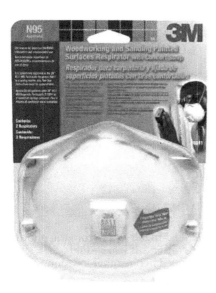

Figure 3-2. *Buy and wear a breathing mask when cutting or drilling MDF.*

You might find it of interest to know that there are many organizations and groups that are pushing formaldehyde-free MDF. It can be a bit difficult to find, but if this is a major concern for you, we encourage you to do some research and locate a vendor who can provide you with formaldehyde-free MDF material.

Label Parts

Take a look at Figure 3-3. This figure shows all the MDF pieces for the CNC machine except for the two sheets of 2'×4' MDF that will make up the CNC's tabletop (x-axis).

Figure 3-3. *It's easier to find what you need when you label each of the MDF pieces.*

You can see that each piece of MDF has been given a letter. It's a little difficult to tell from the photo, but the letter is written on the front and back of each piece on opposite corners. (If you should accidentally trim a piece along any one edge, the piece should still be labeled in one other location.)

NOTE In later chapters, we'll use letters to reference parts, instead of their actual names. You'll find the assigned letters in the Part Letter(s) column of the CNC machine spreadsheet and on each of the parts in the Parts Layout PDF files. Both of these files are available for download at www.buildyourcnc.com/book.aspx.

Don't stop there, however. You're going to find as you build your CNC machine that it also helps to label the front and back of each piece (F and B, written somewhere obvious). This will come in handy once you start assembling your CNC machine and can't remember which side of a piece of MDF faces outward or inward; sometimes it doesn't matter and sometimes it does.

MDF Sheets and Sizes

When you go to buy your MDF for your CNC machine, you'll likely find the MDF sold in 4'×8' sheets (3/4" thick). (To make things even more confusing, some MDF sheets are a little oversized, so you may find the measurements are closer to 49"×97"—an extra 1" on each side.)

The CNC machine you'll be building calls for using four sheets of 2'×4' MDF. You have two options here:

1. Buy a single 4'×8' sheet and trim it yourself into four 2'×4' sheets.

2. Ask the seller to cut the 4'×8' sheet into four 2'×4' sheets.

Most hardware/lumber stores do not like to cut MDF due to the dust it creates, but some will—it never hurts to ask. But make certain to ask them to trim them to 2'×4' sizes, especially if the single sheets come with a little extra material on the edges.

You may luck out and find a seller that sells quarter-size MDF sheets, precut to the 2'×4' size (see Figure 3-4). You may pay a little more per quarter-sheet, however. Trust us—it's probably worth the extra dollar or so per quarter-sheet, and the cuts will likely be extremely accurate and easier to carry (MDF is heavy).

Figure 3-4. *Try to find MDF in quarter-sheets with 2'×4' dimensions.*

The quarter-sheets in Figure 3-4 were priced at $7.95 per sheet when the photo was taken. Multiply that by four sheets and you get approximately $32.00. Compare that to a single 4'×8' sheet sold at the same location for $26.00. Not having to carry and cut that big heavy sheet: Priceless. OK, maybe not—but definitely at least worth the extra $6.00 if you can find the quarter-sheets. Also, keep in mind that pricing will very likely fluctuate from location to location; the prices we quote in the book may not match exactly what you pay.

Limit Your Cuts

Take a look at Figure 3-5. It shows a listing of the parts you'll be cutting along with their dimensions. Notice that some of the parts have identical measurements—either in width or length.

Name of Part	Number of Pieces	Length	Width	Part Letter(s)
Y Axis Back Support	2	10 1/16"	2"	A and B
Z Axis Rail	1	10 1/16"	4"	F
X Table End Feet	2	16"	6 25/32"	T and U
Gantry Sides	2	17 3/4"	7"	Q and R
Gantry Bottom Support	1	2"	4"	E
Y Axis Rail Support Front Reinforcement	1	23 11/16"	6"	S
Gantry Bottom	1	26 11/16"	7"	P
Y Axis Back Support	1	26 11/16"	8"	O
Motor Mounts	6	3 7/16"	2 1/2"	G, H, I, J, K, L
Table	2	49"	24"	
Z Axis Back Support	2	6 1/16"	1 1/2"	M and N
Router Base	1	6 1/16"	8"	V
Y Axis Linear Bearing Supports	2	6 7/8"	4"	C and D
Z-Axis Bearing Supports	2	8 13/32"	7"	W and X

Figure 3-5. *Try to group your parts so you'll make as few cuts as possible.*

You can download the entire listing in spreadsheet format at www.buildyourcnc.com/book.aspx. After downloading the file, open it and you'll see some of the part names match up with Figure 3-6.

Figure 3-6 shows one of the two PDF files that contain the measurements and part layouts for two of the 2'×4' MDF sheets. These part layouts are designed to help you make fewer cuts, as well as to help make certain that paired parts (such as the two gantry side pieces) are identical in size because they're cut together along one or more edges.

You can download the part layout files at www.buildyourcnc.com/book.aspx. Keep in mind that there are only two of them because two of the four 2'×4' sheets will be used to make the CNC machine's tabletop and require no cutting.

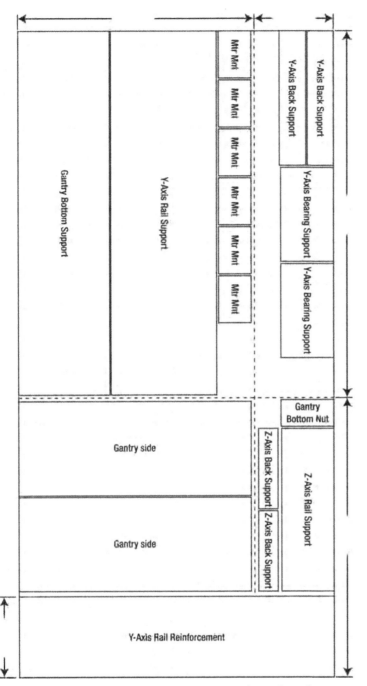

Figure 3-6. *The CNC machine parts are grouped to reduce the amount of cuts you'll need to make.*

Time Your Cuts

This may not make sense now, but trust us—try and cut only the parts you need, when you need them. Look back at Figure 3-6. The two parts in the lower-left corner are the Y-Axis Rail Support and the Gantry Bottom Support. These two parts need to be identical in length, but this length won't really be known until you reach the point where you'll assemble the Y-Axis Gantry. See the dotted line running vertically to the right of these two parts? That cut can be made to separate these two parts from the rest of the MDF sheet, but it's not the final cut that's indicated by the solid line on the right side of both parts.

Download and examine carefully the part layout files mentioned in the previous section. You should see that each 2'×4' sheet is further divided into sections that can be set aside until the parts they will provide are needed.

TIP If you follow our advice from Chapter 2 and read through the entire book, especially those chapters focused on the cutting and assembly of the CNC machine, you should be able to locate the parts you'll need first—the Z-Axis Rail Support and Z-Axis Bearing Supports, for example. Those parts are conveniently located in the part layout sheets for first cuts.

But we hear you yelling "Hey, Figure 3-3 shows all the parts already cut out!" Yes, it does. But what you can't see in Figure 3-3 is the frustration we encountered when we figured out (later, after this photo was taken) that we really shouldn't have cut out the Y-Axis Rail Support and Gantry Bottom Support until we needed them!

We'd also like to share with you another one of our big mistakes in the hopes that you avoid it. You'll be using two 2'×4' sheets of MDF to make the CNC machine tabletop. These two pieces will be chamfered (see Chapter 7) and bolted together, and should be flush along all the edges. Well, in our eagerness to get ahead, we chamfered both sheets, measured and drilled the top sheet, squared it up, clamped it along the left and right sides of the bottom sheet, and drilled the holes. And guess what happened? The bottom sheet, which we assumed was exactly 2'×4', was actually 1/4" wider than the top sheet, but we didn't notice that until we bolted the two sheets together—the chamfered edges didn't match up! (Not a huge mistake, but it slowed us down since we had to unbolt the sheets, trim the bottom sheet, rechamfer it, etc.)

The lesson to be learned here is to measure all your parts and make certain they match up with the dimensions found in the actual CNC machine plans.

It never hurts to double or even triple-check your measurements against the part layout sheets and the actual CNC machine plans!

Encouragement

You can do it! OK, maybe you need a little more than that. If so, go back and watch the videos of Patrick at www.buildyourcnc.com and keep this in mind—the majority of that CNC machine is being built by Patrick using nothing but a mitre box and saw, a portable drill, and some screwdrivers and wrenches. Even more encouraging, Patrick had no previous CNC machine–building experience! A few "exotic" tools, such as a router, will be needed at some point, but the reality is that this project is totally within your reach.

Don't view this project as a bunch of hurdles to overcome. And don't treat it as a race. Take your time. When you get tired, take a break. When you get frustrated, take a break. Just remember that others have built this thing and so can you!

But let's be realistic for a moment and talk about some of the bumps you're likely to encounter on this trip. Did you cut a piece incorrectly? Don't freak out. First, take a short break and then come back. Take a look at the part layout sheets and you're likely to see almost half of a single 2'×4' MDF sheet unused. This isn't waste material—it's extra MDF for when you make a mistake. And you will—we're not going to lie to you. Whether you cut or drill a piece wrong, just keep in mind that you've got some extra MDF there. The only thing your mistake has cost you is a little bit of time. (And if you use up all your spare MDF, just remember that MDF isn't that expensive in the big picture.)

Did you drill a piece in the wrong spot? Again, don't get angry. You may find that you can drill the correct hole without any problems other than having that "bad" hole visible on your CNC machine. If you're a perfectionist, go cut another piece if it's going to bother you. If you don't care if your CNC machine has an extra hole in a part here or there, then just roll with it and try to be more careful measuring and drilling the next piece.

Do the pieces not "square up" properly or match exactly the way they need to? Take a step back and examine the problem. Did you accidentally flip a part, and maybe the drilled holes aren't lining up because you've got the back of the piece facing outward instead of the front? Can you loosen up some of the bolts you've already attached and fix the problem before tightening down all the nuts? We promise you this— if you've cut the parts to the correct dimensions and drilled the holes properly, you're likely facing something simple and easily fixed.

Throughout the project, be patient. Give yourself three strikes per work period—if you make a mistake, slow down and fix it before moving forward. If you make a second mistake, it's time to really pay attention to your work. If you find you make a third mistake, it's time to call it quits for the day and come back later. Trust us—your CNC machine isn't going anywhere.

What's Next?

Chapter 4 will provide you with some of the information on how your CNC machine will actually move and cut parts. It's not done using any extremely complicated devices. If you thought your CNC machine would be using complex hardware to move forward, backward, side to side, and up and down, you're in for a pleasant surprise when you see the simple solution you'll be building and implementing on your own machine.

Movement Using Rails

Your CNC machine isn't a static device. It will be using motors and other hardware to move up, down, left, right, back, and forth. This movement will need to be as smooth as possible. Imagine trying to cut or drill something by hand while someone behind you keeps pushing, pulling, and jostling you. It would be extremely difficult. Like most people, you'd do your best cutting and drilling without any outside interference.

Your CNC machine requires the same environment. You can control certain things such as the surface that you place your working CNC machine on; flat and nonmoving is best. (This rules out sitting it on top of your washing machine during the rinse cycle.) And when it's operating, you can ensure the best results by not letting anyone or anything touch the device (including pets and children).

But your CNC machine is also going to require that its own parts are moving smoothly. Back in Chapter 1 we talked about how the machine's cutting and drilling tool (most likely a router) will be moving up and down along the z-axis, side to side on the y-axis, and forward and backward along the x-axis. For the best drilling and cutting results, these movements need to be extremely smooth. The MDF material you'll be building your CNC machine out of is smooth, but not as much as your machine will really need. In order to get the best movement along all three axes, your CNC machine is going to use an inexpensive solution that's also extremely smooth and accurate. Here's how it works.

Bearing-Rail Assembly

Follow along by carefully examining this section's figures, and we'll explain how this solution will work so the title of this section will become clearer. (Don't worry about any MDF pieces or other stuff in the pictures; for now, just pay close attention to those pieces we point out in the figures.)

The two items in Figure 4-1 are 4" lengths of aluminum angled rail. The rail consists of two 1/8"-thick walls that meet at a 90 degree angle (right angle). Rail width is measured on the outside wall from the outside edge to the edge where the two walls meet. You can typically purchase this angled rail in lengths of 4', 6', and 8'. (Rail is easily cut with a hand saw, but feel free to use whatever method you prefer for cutting the rail to the proper lengths needed for your CNC machine. You'll also be told the proper lengths to cut your rails in later chapters.)

Figure 4-1. *Two pieces of aluminum angled rail*

Figure 4-2 provides more detail on how to find the proper type of aluminum angled rail; even though rail is often labeled with the proper measurements, don't trust the sticker. Measure it yourself to make certain you're buying the proper thickness and width of rail.

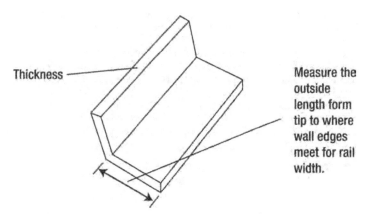

Thickness

Measure the outside length form tip to where wall edges meet for rail width.

Figure 4-2. *The angled rail's width is measured from the outer wall edge to the inner wall edge.*

For your CNC machine, you'll be purchasing two different widths of rail: 3/4" and 1 1/4". The thicknesses on both rail widths will be 1/8". The rails shown in Figure 4-1 are 3/4" wide and 1/8" thick.

Now take a look at Figure 4-3. This figure shows a few of the pieces of hardware that you'll be purchasing—bearing, bolt, and nut. It also shows the three items assembled. (You'll be given specifics on sizes and quantities later in the book.)

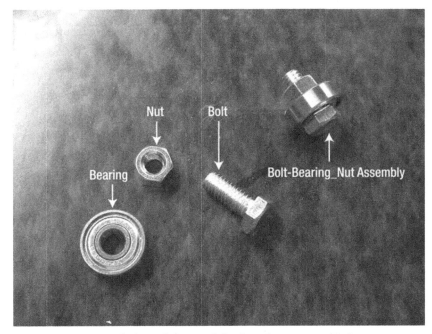

Figure 4-3. *Bearing, bolt, and nut, and the three items assembled*

Just in case you're wondering, the bearing shown in Figure 4-3 is the same type of bearing you'll find used in skates!

Before we continue, we want to give you a preview of what a final bearing-rail assembly (BRA) will look like. Figure 4-4 shows a piece of angled rail with four of the little bolt-bearing-nut assemblies mounted to it.

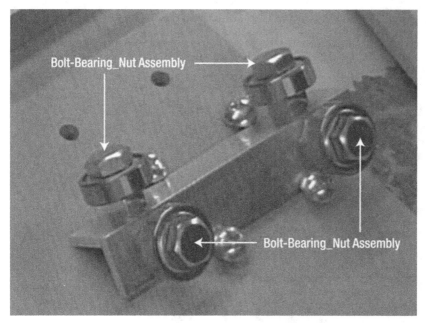

Figure 4-4. *A BRA—one rail and four sets of bolt-bearing-nut assemblies*

Now let's take a step back; you need to know how to take four of the bolt-bearing-nut assemblies and attach them to the rail to make a BRA.

After the bearing, bolt, and nut are assembled, they must somehow be attached to the piece of rail you've cut. Attaching will involve three steps: drilling a pilot hole, drilling a 17/64" hole, and *tapping* the hole (creating threads in the hole for a bolt to be screwed into).

First, you're going to drill pilot holes in the rail. Let's talk about pilot holes. You're going to hear that used in many places in the book, so this is a good time to explain what they are and why they're important.

Drilling a pilot hole could also be considered *predrilling* because you're going to use a very small-diameter drill bit to drill a starting hole. Take a look at Figure 4-5 and you'll see a small piece of metal—about 1/16" thick—with a single hole in it.

This little piece of metal serves two purposes. First, we'll place it over the rail where we wish to drill a hole. This will allow us to drill the hole in the same location on every piece of angled rail that we've cut. Figure 4-6 shows how the metal is being used as a template for drilling a hole in our angled rail. (You can purchase a small scrap of 1/16"-thick metal like this at a local hardware store cheaply.)

The small hole will allow you to more accurately drill a final hole in the aluminum. Notice in Figure 4-6 that the metal piece has been pushed completely up against the inner edge of the rail (where the two walls meet) and is flush with the end of the rail as well.

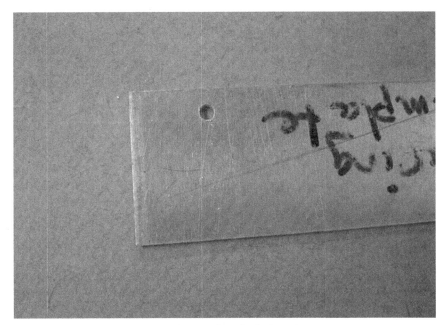

Figure 4-5. *A piece of metal with a single pilot hole in it*

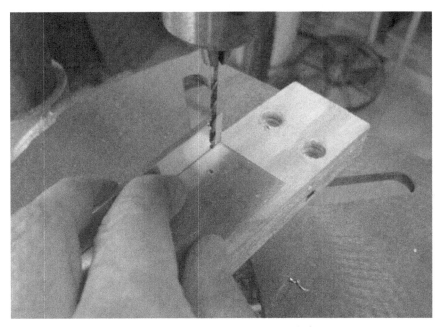

Figure 4-6. *Drilling a pilot hole into an aluminum angled rail piece*

Figure 4-7 shows the measurements for where the pilot hole was drilled in the metal template we used. You can download a PDF file containing this image and measurements at www.buildyourcnc.com/book.aspx.

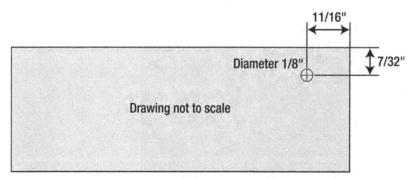

Figure 4-7. *The metal template used to accurately drill pilot holes*

We know that the rail will require four sets of the bolt-bearing-nut assembly, so you can simply flip the metal template over and drill a matching pilot hole on the opposite side of the rail's wall. Then drill two more holes on the rail's other wall. Figure 4-8 shows a rail piece with all four pilot holes drilled.

Figure 4-8. *Four pilot holes drilled into a rail*

The small pilot hole will allow you to center a larger drill bit more accurately. Most drill bits have a sharp tip on their end that will fit inside the small pilot hole—take advantage of this fact and your drilled holes will be more accurately placed.

NOTE We use the same pilot hole method for drilling in MDF as well. You don't have to use pilot holes, but trust us—spending the extra time to drill pilot holes is well worth it and will help ensure your hole is drilled exactly where you measured and marked it.

The next thing you'll need to do is enlarge your pilot holes. For all the rails you will make for your CNC machine, you'll be drilling 17/64" diameter holes. You can see in Figure 4-8 the larger-diameter drill bit that will be used to make these holes. The 17/64" drill bit typically comes with the package that includes the tap and is usually labeled as a drill bit and tap for 5/16" screws.

Why are we drilling using a 17/64" bit and not a 5/16" bit for the 5/16" bolt that will screw into the threads? The hole will be a little smaller than 5/16" because you're going to cut grooves (threads) into it in the next step. If you drill an actual 5/16" hole, the 5/16" bolt would just fall through. The little extra bit of aluminum left over (3/64") with the 17/64" bit will leave just enough material for the threads to be cut so the bolt can screw into the hole.

Figure 4-8 shows that we've clamped down the aluminum rail for drilling using the 17/64" drill bit. Figure 4-9 shows the rail after the holes have been drilled.

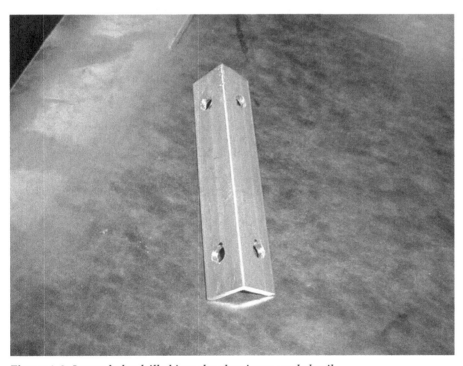

Figure 4-9. *Larger holes drilled into the aluminum angled rail*

Finally, in order to insert the bolt-bearing-nut assemblies, we need to create a thread inside the 17/64" holes for the bolts to be screwed into. To do this, we use a tool called a *tap*. You can see this tool in Figure 4-10.

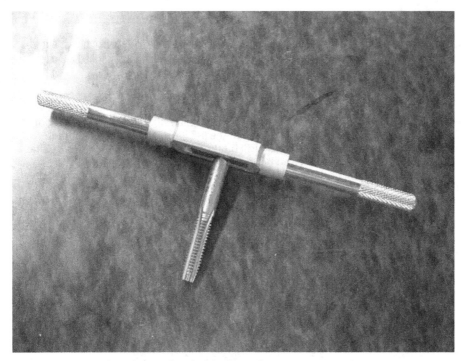

Figure 4-10. *Cut threads into the aluminum rail using a tap.*

The t-shaped tool is used to cut threads into the 17/64" holes. You place the tap bit into the hole, making sure that the tap is as straight and vertical as possible with respect to the hole. You then twist the tap clockwise and the tap bit will cut into the aluminum edges of the hole. Figure 4-11 shows the tap cutting threads into a hole.

Keep twisting—you'll feel resistance, but it's shouldn't be extremely difficult to twist. At all times try to keep the tap bit going straight into the hole, not at an angle. Figure 4-12 shows the tap bit over halfway through the hole.

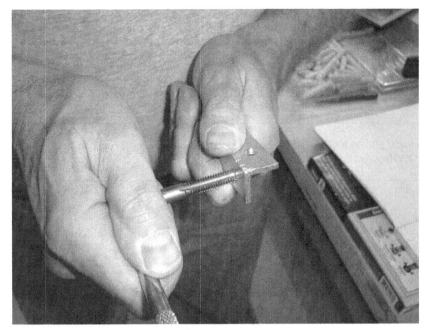

Figure 4-11. *Cutting threads into one of the holes in the aluminum rail*

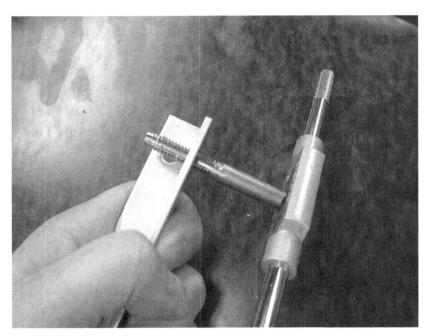

Figure 4-12. *The tap bit over halfway through the hole*

When the tap bit gets about halfway into the hole, it will become extremely easy to twist and very loose. Give it a few more twists and then spin the tap counterclockwise to remove it completely from the hole. Look inside and you should see small threads that will allow a bolt to be screwed in. Do this three more times per rail and you'll end up with a piece of rail with four threaded holes. Screw in four of the bolt-bearing-nut assemblies and you'll end up with a rail like the one shown earlier in Figure 4-4. Tighten the bolts down lightly with a wrench—do not apply too much force or the threads can become damaged. We'll refer to this final piece of rail with four bolt-bearing-nut assemblies as a BRA.

NOTE Videos for using the tap and putting together a BRA can be found at `www.buildyourcnc.com/book.aspx` —click the Chapter 4 Videos link to view them.

Riding the Rail

Now it's time to see how your CNC will move smoothly. Figure 4-13 shows an unused piece of rail and a completed BRAs for short.

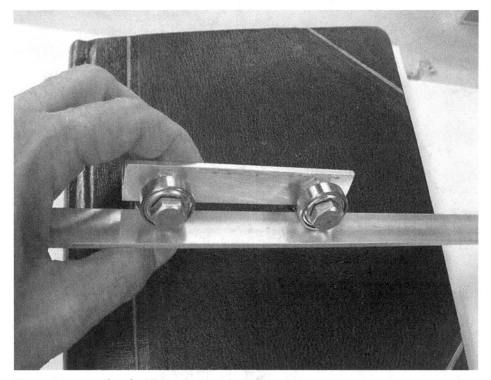

Figure 4-13. *A rail and a BRA*

Place the completed BRA so all four bearings rest against the flat wall surfaces of the rail. Slide the BRA back and forth along the rail. Smooth, isn't it? This is one of two mechanisms you'll be using to give your CNC machine smooth and accurate movement. (You'll learn about the second method in Chapter 6.)

Tips and Advice

Do you see the gap between the BRA and the rail it rides on in Figure 4-13? If the gap is too small, the two pieces will rub against one another and can create resistance. This will cause your CNC machine's movements to be less than smooth. To avoid this problem, it's important that the spinning bearings be a certain distance "up and away" from the rail it rides.

Fortunately, you won't have to do any trial-and-error drilling for your BRAs. If you download, print, and use the template we mentioned earlier in the chapter, you can mark and drill all your rails in the proper spot so the bearings will spin smoothly and the gap will be sufficient in size to prevent rubbing.

It's also very important when tapping the 17/64" holes that you twist the tapping tool straight into the hole. Go slow—twist, stop, and rotate your hands around so you can see all angles. Is the tapping bit going into the hole at any angle other than vertical (90 degrees)? If so, apply some pressure and force the tapping tool back into proper vertical alignment. You might want to cut a scrap piece of rail, drill a few pilot holes and 17/64" holes, and practice tapping—it's well worth the time so you can create the perfect BRAs to ride the rails of your CNC machine.

You'll be building a total of six BRAs—two for the x-axis, two for the y-axis, and two for the z-axis. Not all the BRAs will be the same length, but the method for creating them is identical—measure and mark where the four holes will be drilled, drill four pilot holes, drill four 17/64" holes, and tap four holes.

Trust us—creating the BRAs is not hard. Just take your time, double-check your marks, drill those pilot holes, and practice your tapping technique on some scrap. We'll tell you in later chapters exactly when it will be time to build some BRAs and how many. You'll do great.

What's Next?

Next, in Chapter 5, you're going to learn about two possible methods for connecting pieces of MDF together. You'll probably be surprised at how easy each method is to use and how strong the connection between the pieces is when it's done.

Joining Methods

If the title of the chapter seems strange, let us explain. The CNC machine you'll be building is made up of 26 varying-sized pieces of MDF material. Those pieces have to somehow be connected in such a way that the machine doesn't fall apart. Keep in mind that your CNC machine will be moving—the z-axis will be moving up and down and side to side on both the y-axis and the x-axis. Often these movements will start and stop quickly. Your CNC machine must be able to withstand the twisting and bending and other stresses required for its operation. Wood glue simply won't do, and the mechanical way of fastening has the added benefit of disassembly. This is a machine, so we're going to treat it like a machine.

This short chapter is going to introduce you to two methods for connecting one MDF piece to another. Both methods create strong connections that will be difficult to break, and both methods use the same tools. We'll leave it to you to decide which method you'd like to use, but we favor one in particular that we'll discuss shortly.

Keep in mind as you read this chapter that almost all the pieces of MDF that you will cut and drill will connect using one of these methods. Pick one and become familiar with it. Listen to the tips and advice we provide, and your CNC machine will be a rugged and durable device for years to come, with minimal maintenance.

Two Pieces of MDF

Before we show you the two different methods, we want to explain how the two methods work by using some simple figures and photos. First, take a look at Figure 5-1; it shows two pieces of MDF.

For most of your CNC machine assembly, you will be joining MDF pieces at right angles. Figure 5-2 shows how the two pieces of MDF will connect.

Figure 5-1. *Two pieces of MDF with no holes*

Figure 5-2. *Two pieces of MDF at right angles to one another*

The pieces will be connected using a series of drilled holes; MDF piece A will have a single hole drilled into it, as shown in Figure 5-3. MDF piece B will have two holes drilled into it, also as shown in Figure 5-3. Don't worry right now about where to drill the holes; you'll get that information in specific chapters when you begin to cut, drill, and build your machine.

CAUTION Part A and Part B in these figures are not the same Parts A and B for your actual CNC machine; these are waste pieces of MDF and are only used to demonstrate two methods for connecting pieces.

Figure 5-3. *Holes will be drilled in each piece of MDF—one hole in Part A and two holes in Part B.*

The hole size drilled in Part A matches the hole size drilled on the edge of Part B. The second hole drilled into the surface of Part B will be of a different and larger size.

When you're reading the CNC machine plans, you'll be told what size holes to drill, but in most instances where you're connecting two pieces of MDF at right angles, the matching holes will be of 1/4" size (use a 1/4" drill bit). The surface-drilled hole (on Part B) will be of a larger size and will depend on whether you use method 1 or method 2 for joining MDF pieces at right angles.

Now that you understand how these two pieces are to be connected, let's look in detail at the two methods available. Note that in each method, a different-sized drill bit will be used to drill the surface hole on Part B.

Method 1: Cross Dowels

Take a look at the hardware components in Figure 5-4. There are two bolts—the length of the bolt you use to connect MDF may vary, but in this instance, it is 2" in length. The smaller components are often referred to as either *cross dowels* or *barrel nuts* (and they're also occasionally called *knockdown nuts*). When visiting a hardware store, if you choose to use method 1, be sure to mention these names to the salesperson if you can't find them. For purposes of this book, we're going to use the term *cross dowel* when referring to this part.

Figure 5-4. *Two bolts and two cross dowels (barrel nuts)*

TIP Cross dowels can sometimes be slightly expensive when purchased individually. You'll be using over 50 of these as you build your CNC machine, so buying them over the Internet may provide you a lower cost per piece.

If you look closely at the cross dowel, you'll see that one end contains a slot—you'll be using a slot screwdriver with this component. Notice also that the cross dowel has a hole through its main body and the hole is threaded—meaning a bolt with threads can be screwed into it. Figure 5-5 shows the bolt screwed into the cross dowel—notice that the slot on the cross dowel is parallel to the bolt. (In other words, the slot points in the same direction that the bolt points.) You will use a slot screwdriver to orient the slot on the cross dowel to match up with the insertion of the bolt.

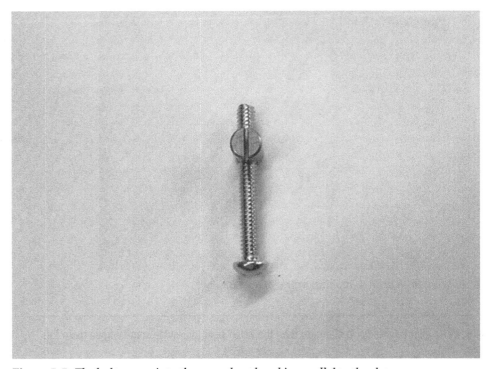

Figure 5-5. *The bolt screws into the cross dowel and is parallel to the slot.*

The cross dowel has a diameter of almost 7/16" and will be inserted into the hole drilled on the surface of Part B. This will require drilling a 7/16" hole in the surface of Part B, as shown in Figure 5-6.

Figure 5-6. *A 7/16" hole has been drilled into the surface of Part B.*

NOTE The cross dowel is slightly smaller in diameter than the 7/16" hole to provide some wiggle room for fitting it into the hole.

Notice that the holes on Part B are aligned. You can see a pencil line in Figure 5-6 that runs from the bottom of the 7/16" hole on Part B; this line is useful for drilling the 1/4" hole into the edge of Part B. If you stick the bolt into the 1/4" hole on the edge of Part B, you should be able to view it by looking down into the hole drilled on the surface, as shown in Figure 5-7.

After removing the bolt, we next take Part A and place the 1/4" hole drilled on its surface over the 1/4" hole drilled into the side of Part B, as shown in Figure 5-8. We also insert the cross dowel into the surface hole on Part B with the slot facing up so we can see it—this is also shown in Figure 5-8.

Figure 5-7. *If drilled properly, the inserted bolt will be visible when you look into the surface hole.*

Figure 5-8. *Align the 1/4" holes on Parts A and B and insert the cross dowel into the surface hole.*

Use a slot screwdriver to twist the cross dowel so the slot runs parallel to the 1/4" drilled holes in Parts A and B. Now we're ready to insert the bolt. This may seem a little tricky at first, but it gets easier as you do more of them—practice makes perfect!

Slowly insert the 1/4"-diameter bolt into the hole on Part A. Push it through into Part B. Your goal is to have the bolt find the hole in the cross dowel and finger thread the bolt into the cross dowel. While doing this, you'll be using a slot screwdriver to make sure that the slot on the cross dowel remains parallel to the bolt so it can be threaded properly.

The typical mistakes made here are pushing the cross dowel too deep into the surface hole or not deep enough. Either way will keep the bolt from being screwed in properly to the cross dowel. Keep at it—you'll eventually find that finger tightening the bolt becomes more difficult and requires a Philips head screwdriver (manual or electric) to tighten the bolt completely.

Notice in Figure 5-9 that the edge of Part A is flush (or level) with the surface of Part B; this is done by lining up the edges just before completely tightening down the bolt. If you've already tightened the bolt down, loosen it just a bit, align the edges properly, and retighten.

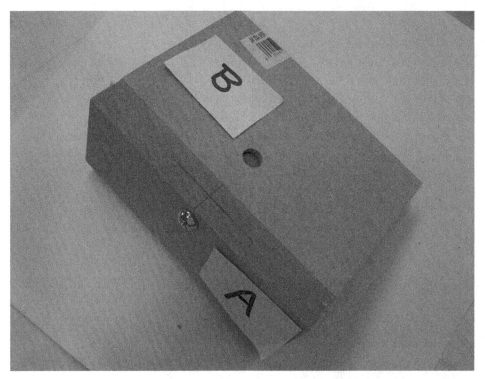

Figure 5-9. *Try to align the edges before completely tightening down the nut and cross dowel.*

Congratulations. You now know how to connect two pieces of MDF using the cross dowel method. Now let's examine method 2.

Method 2: Bolt, Washer, and Nut

You'll be happy to know that method 2 is almost identical to method 1 except that you won't be using a cross dowel. Instead, you'll drill a slightly bigger hole on the surface of Part B (at the same location where you drilled for the cross dowel) and insert a standard nut that the bolt will screw into and tighten to secure the MDF.

Take a look at Figure 5-10 and you'll see the same setup as before; we have two pieces of MDF, Part A and Part B, that need to be secured at a right angle.

Figure 5-10. *Parts A and B will be secured using a single bolt and nut.*

You'll drill the same 1/4" holes in Part A and Part B for insertion of the bolt; in Figure 5-10, this is a basic 2" bolt (1/4" diameter) with a matching nut. The larger drilled hole on the surface of Part B will be 3/4" diameter (use a 3/4" drill bit). You will insert the nut sideways into the 3/4" hole.

Place Part A and Part B so the edges are flush (or the parts are matched up at the edges), as shown in Figure 5-11. Insert the bolt into the drilled hole in Part A and then into Part B. Hold the nut sideways, as shown in Figure 5-11, and hand-thread the bolt into the nut. Use a Phillips-head screwdriver to complete the tightening of the bolt.

Figure 5-11. *Parts A and B are connected by securing the bolt with the nut.*

It's an alternative to using cross dowels that works well. One potential drawback to this method, however, is that over time the bolts will need to be tightened more often; they will tend to loosen as the CNC machine vibrates and moves.

Which Method Is Best?

We favor the cross dowel method, but cross dowels do cost a little more per piece. Both methods will allow you to tighten down the connected MDF pieces securely, but cross dowels seem to be less likely to loosen over time and require only tightening of the bolt.

TIP You might also consider using lock washers or lock nuts. Lock washers are used to prevent bolts and nuts from loosening. You put a lock washer on just before screwing on the nut. Lock nuts have a nylon insert inside them that also helps prevent loosening of its bolt.

Pick a method that works best for you—base it on the skills and tools you have available. You'll get plenty of practice in later chapters as you begin building your own CNC machine, but feel free to try both methods on some scrap MDF pieces and see which method you like best. You may find that method 2 is easier for assembly because you don't have to fumble with aligning the cross dowels—that's fine. You may like method 1 because it doesn't require you to find a wrench to hold the nut for tightening (or hurt your fingers).

Building a Jig to Drill

Now that you've seen the two methods used to connect two pieces of MDF, we'd like to offer you one additional tip that may make drilling easier. In our experience, we have found that a simple jig (guide) may help in aligning the holes drilled to connect pieces of MDF. With the large amount of right angle joining necessary, a jig can be extremely useful. Once you drill the first hole, finding the intersecting hole location with the proper alignment to the first hole can be trying without some help. A jig will help in two main ways. First, a jig can help with drilling in the proper orientation (drilling a straight hole). Second, the jig will insure the proper location for the intersecting hole and for the hole into the edge of the piece with consistency.

You can build your jig from a small piece of scrap MDF; make certain it's constructed from the same thickness of MDF as you're using for your CNC machine—3/4". Draw a straight line down the center of the piece and then cut the piece in two, as shown in Figure 5-12.

Figure 5-12. *Cut a scrap piece of MDF to create your own drilling jig.*

In Figure 5-13, the jig is constructed by gluing the two scrap pieces of wood joined at a right angle and the measurements where the holes need to be positioned. Don't drill the holes until after you've cut the two pieces; measure the location of the holes on the line that was drawn down the center of the pieces.

Figure 5-13. *Apply some wood glue to the edge of one piece of scrap MDF.*

In Figure 5-14 you can see how the two pieces are joined. After you've selected the method you prefer to use (method 1 or method 2), you can place your jig over the MDF you'll be drilling to drill holes exactly where you want them.

Figure 5-14. *Clamp the jig over the piece to be drilled and drill precisely placed holes every time.*

If you place the jig over the piece of MDF to be drilled, the two pilot holes shown in Figure 5-14 will place those holes at the standard distances from the edges of the MDF indicated in the CNC plans.

What's Next?

In Chapter 4 you learned how the BRAs (Bearing-Rail Assemblies) allow your CNC to move smoothly along pieces of rail. In Chapter 6 you're going to learn the second half of what's required for CNC movement. The BRAs won't roll themselves, so something has to provide them with the proper force (and the proper direction) for movement. Luckily, it's an extremely simple and elegant solution that you'll find easy to implement.

The Electronics

Your CNC machine consists of two parts: the MDF (and the bolts, bearing, nuts, and other hardware) and the electronics. The majority of the chapters in this book will cover the building of the CNC machine—cutting, drilling, and bolting together various MDF pieces to make the physical frame of the machine. The other part of the machine is the electronics. Three motors will be used to move the router and allow your CNC machine to do the job you want it to perform. These motors are controlled using some additional electronics, which we'll talk about shortly. It is these electronics that will connect to your computer and receive instructions from the special software you'll be using to tell the machine where to cut, drill, and perform other actions.

There are numerous options available for the electronics used in your CNC machine. This means it is simply impossible to cover every type of electronic device and how to incorporate it into your machine. Because of this, we are going to be using a readily available set of electronics that you can purchase directly from one of the authors (Patrick) or from a couple of vendors. We'll give you exact part names (and/or numbers) so you'll be able to purchase the exact hardware used in the instructions provided in this chapter.

If you choose to purchase different electronics for your CNC machine, be sure to read through all the documentation available. You may find that some electronics use different wire colors or different labels for various parts, but with some careful examination of the documentation and schematics, you may be able to figure out how the differences match up to our instructions.

CAUTION The electronics are sensitive to improper wiring and most of them aren't going to survive when improperly provided the wrong amount of voltage. Use a multimeter (volt meter) whenever you can to verify voltages and always read any documentation provided with your electronics for proper wiring and configuration.

The Required Components

You're going to need to purchase the following nine items (in addition to the wire, solder, and other small electronics components mentioned later in this section):

```
Three stepper motors
Three stepper motor drivers
One breakout board
One power supply
One 5V DC transformer 400mA (6V is OK)
```

Three stepper motors will be required, one per axis. Each motor comes with a wire bundle containing eight wires of different colors. The stepper motor we will be using in this chapter is model KL23H286-20-8B from www.kelinginc.net. You can see one of the stepper motors in Figure 6-1.

Figure 6-1. *A single stepper motor with wire bundle*

You'll also need to purchase three stepper motor drivers; each motor will receive electrical signals from a stepper motor driver that tells the motor how fast to spin, as well as controls the direction of spin of the motor's shaft. For our CNC machine, we're using the Keling KL4030 stepper motor driver shown in Figure 6-2.

Finally, you'll need to purchase a single breakout board. The breakout board connects to your computer via the parallel port. The special CNC software installed on your computer sends signals over a standard parallel printer cable to the breakout board. Each stepper motor driver also connects to the breakout board. When wired properly, the computer software will be able to control the individual stepper motors (sending signals to the stepper motor drivers through the breakout board). The breakout board we're using is the Keling C10, and looks like the one shown in Figure 6-3.

Figure 6-2. *One stepper motor driver is required for each stepper motor.*

Figure 6-3. *The breakout board connects to your computer and all three stepper motor drivers.*

To provide power for the stepper motors and the cooling fan, you'll need to purchase a power supply. We're using Keling's KL-350-36 36V/8.8, as shown in Figure 6-4.

Figure 6-4. *The power supply provides the proper voltage to all the electronics.*

In addition to these nine items, you'll also need the following items:

50 feet of black 18-gauge wire

50 feet of red 18-gauge wire

10 feet of green 18-gauge wire

10 feet of white 18-gauge wire

TIP You can save some money on wire by purchasing 50 feet of AWG 4-conductor PVC cable; this cable consists of an outer sheath covering four 18-gauge wires inside. You can slice open the sheath and pull out the four wires to trim them to whatever length you need. If you're not able to locate this cable, contact www.acksupply.com to order it. This may very well be cheaper than purchasing the four colors above in various lengths.

Solder gun and solder

Wire strippers

Multimeter

Plastic wire conduit (not required, but extremely helpful—we used 1" diameter)

8' extension cord for power supply (the kind with three prongs—the smaller of the two rectangular-shaped prongs is the hot wire, the larger is common/neutral, and the rounded one is for the ground)

Shrink tubing (for covering the solder connections)

120V fan (the kind that can be installed in a computer case to cool it)

Two-prong 5' extension cord (for the 120V fan)

Preparing the Stepper Motor Wires

The first thing you're going to do is separate the wires on each stepper motor. For the KL23H286-20-8B stepper motor, strip about 3/4" from the end of each wire and pair up the following wires by twisting the ends together.

Red/white-striped wire and blue/white-striped wire (RW/BW)

Red wire and blue wire (R/B)

Black/white-striped wire and green/white-striped wire (BkW/GW)

Black wire and green wire (Bk/G)

Figure 6-5 shows one of our motors with the eight wires paired up. In addition to twisting the ends of each pair together, we've also used some electrician's tape to hold the wires together.

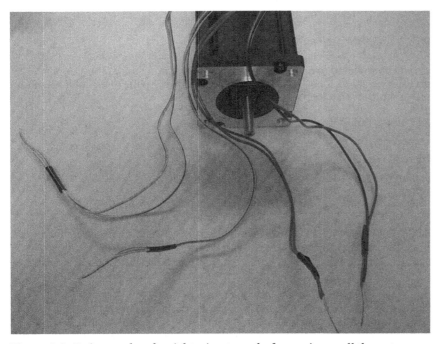

Figure 6-5. *Twist together the eight wires to make four pairs on all three stepper motors.*

Do the same thing for all three stepper motors and then set the motors aside. Next, cut off approximately 8' of wire in four colors (we used black, white, red, and green), as shown in Figure 6-6.

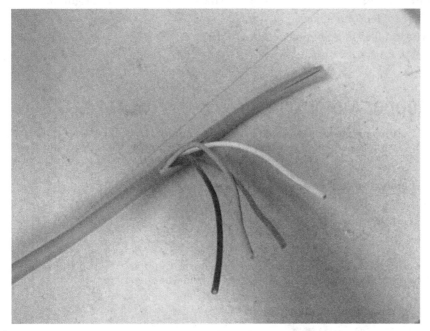

Figure 6-6. *Cut 8' of wire in four colors for each motor.*

Each of these four wires will be soldered to one of the wiring strands coming out of the stepper motors. Try to be consistent with the colors if possible. We chose the following pattern for soldering the 8' lengths of wire to the stepper motor wire pairs:

Black 18-gauge wire soldered to the Bk/G pair

Green 18-gauge wire soldered to the BkW/GW pair

Red 18-gauge wire soldered to the B/R pair

White 18-gauge wire soldered to the RW/BW pair

TIP You don't have to solder the wires together, but soldering will make a more durable connection between the wires. If you choose not to solder, use the proper size wire nuts to twist the wires together.

Figure 6-7 shows one of the wire pairs (B/R) being soldered to the red wire.

After applying the solder, we also used a small piece of heat-shrink tubing to cover the solder. Simply cut a small piece of the tubing, slide it over the soldered connection, and apply a little heat to it (keep moving the lighter back and forth to prevent burning). Figure 6-8 shows us using a lighter to shrink the tubing.

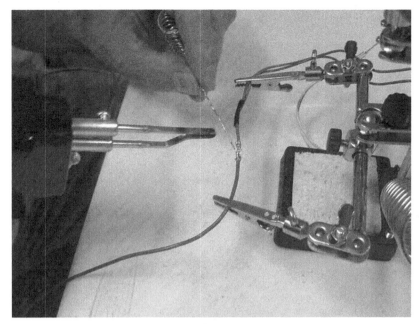

Figure 6-7. *Solder a wire pair to its matching 18-gauge wire.*

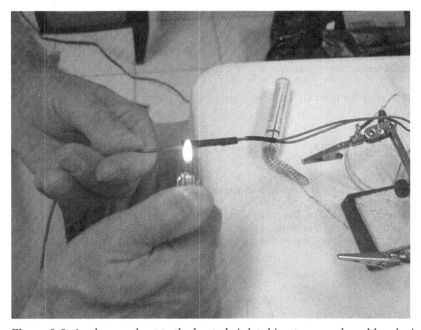

Figure 6-8. *Apply some heat to the heat-shrink tubing to cover the soldered wire.*

Do this for all four pairs on each of the three stepper motors. Figure 6-9 shows one of the motor's wire pairs after soldering the 8' wire strands.

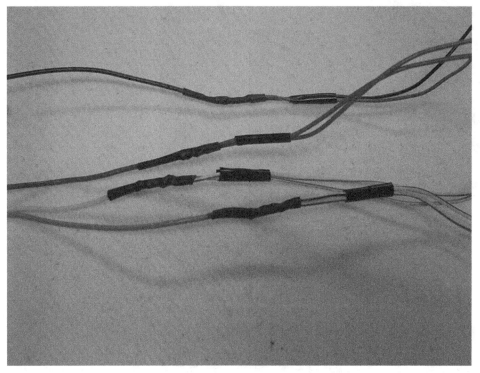

Figure 6-9. *All four wire pairs soldered to their matching 8' wire and covered in shrink tubing*

Go ahead and set the motors aside for now. Up next, we'll prepare the power supply.

Preparing the Power Supply

The power supply does not come with a power cord, so you'll need to cut the female end off of a short extension cord (8' should be enough length) and strip approximately 3/4" from each of the three wires. Figure 6-10 shows the extension cord with the exposed wires.

It's important for you to identify the hot wire, the neutral (or common) wire, and the ground wire. For most extension cords, the ground wire corresponds to the circular metal post, as indicated in Figure 6-10. The hot wire corresponds to the smaller of the two rectangular posts, and the neutral wire corresponds to the larger rectangular post.

Figure 6-10. *Separate the three wires found in a standard extension cord.*

Use your multimeter to determine which wire corresponds to a post. It may be helpful to label the wires with colored tape. As you can see in Figure 6-10, we've also applied a little solder to each wire (called tinning) to keep the small copper strands from fraying.

Grab your power supply. Make sure the extension cord is not plugged in, and connect the wires to the power supply by screwing them down as follows:

Hot wire to L (see Figure 6-11)

Common wire to N (neutral)

Ground wire to the symbol for ground (again, see Figure 6-11)

Plug in the extension cord and use your multimeter to take a voltage reading. Connect the red probe to one of the +V screws and the black probe to one of the –V screws. You should receive a reading of around 36V, as shown in Figure 6-12. Test all three +V screws (and matching –V screws) to make sure approximately 36V is being supplied at each of those terminals.

Figure 6-11. *Connect wires to the power supply by screwing them down to the proper slots.*

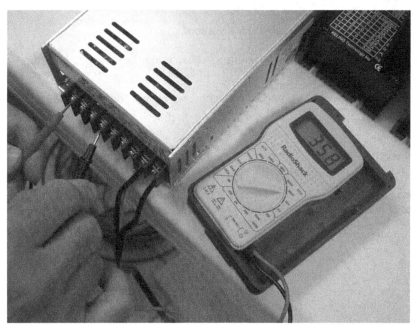

Figure 6-12. *Test the voltage being supplied by your power supply.*

If you're not getting close to 36V (+/−1 volt should be acceptable), unplug the extension cord and check that you've wired up the power supply properly; check to make sure that the proper wires are screwed down on the N, L, and ground terminals. Set the power supply aside and grab your breakout board next.

Preparing the Breakout Board

Next, take your 5V adapter (6V is acceptable) and cut off the end indicated in Figure 6-13. You should find a black wire and a black/white-striped wire. Strip 3/4" off the ends of the two wires.

TIP Strip the ends of the two wires and check the DC voltage using your multimeter. The black probe on the black/white-striped wire and the red probe on the black wire should give a DC voltage around 5 to 6V.

Figure 6-13. *A 5- or 6-volt adapter will power the breakout board.*

Next, screw the two wires into a screw terminal, as shown in Figure 6-14. Screw the hot wire (the black/white-striped wire for our adapter) into one of the screw terminal ports, and screw the neutral wire (black wire for our adapter) into another one of the screw terminal ports.

Figure 6-14. *The screw terminal will help provide power to the breakout board and other devices.*

Next, cut one black piece of wire and one red piece of wire to a length of approximately 10". From this point forward, we'll be trying to use red wires to indicate the hot wires and black wires to indicate the common or neutral wires. Strip these wires on both ends (about 3/4") and use some solder to tin them if you like.

Screw the red wire into the wire terminal opposite to where the hot wire from the 5V adapter is connected (our black/white-striped wire). Screw the black wire into the wire terminal opposite to where the common/neutral wire from the 5V adapter is connected (our black wire).

Grab your breakout board and connect the red wire to the 5V terminal indicated in Figure 6-15. Connect the black wire to the GND terminal (also indicated in Figure 6-15). Finally, cut a 2" piece of red wire, strip both ends, and daisy-chain the wire from the 5V to the ENABLE terminal, as shown in Figure 6-15—simply insert one end into the 5V terminal (it will share the terminal with the hot wire from the wire terminal) and screw down the other end into the ENABLE terminal. Figure 6-16 shows a closeup of the breakout board after the wires are connected.

Figure 6-15. *Wire up the breakout board by connecting the wires from the screw terminal.*

Figure 6-16. *Closeup of the breakout board with wiring from the screw terminal and jumper cable.*

Now grab the three stepper motor drivers and place them as indicated in Figure 6-17.

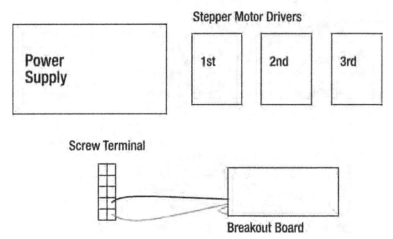

Figure 6-17. *Place the stepper motor drivers so they can be supplied with power.*

TIP You can cut a piece of MDF or plywood for mounting all these electronics so they don't move around. We used a piece of plywood with dimensions of 9"×14" to lay out all the parts shown in Figure 6-17.

Providing Power to the Stepper Motor Drivers

Cut a piece of red wire (hot) and strip both ends (3/4"). The length of this wire will depend on the distance from the hot (black/white-striped) terminal on the screw terminal to the PUL+ (+5V) terminal on the first stepper motor driver. Cut a 2" piece of red wire, strip both ends, and daisy-chain it from the PUL+ (+5V) terminal on the first stepper motor driver to the DIR+ (+5V) terminal. You can see this in Figure 6-18.

Cut another piece of red wire—this wire will be daisy-chained from the DIR+ (+5V) terminal on the first stepper motor driver to the PUL+ (+5V) terminal on the second stepper motor driver. Cut another 2" piece of red wire, strip the ends, and daisy chain it from the PUL+ (+5V) terminal on the second stepper motor driver to the DIR+ (+5V) terminal.

Finally, cut another piece of red wire—this wire will be daisy-chained from the DIR+ (+5V) terminal on the second stepper motor driver to the PUL+ (+5V) terminal on the third stepper motor driver. Cut another 2" piece of red wire, strip the ends, and daisy-chain it from the PUL+ (+5V) terminal on the third stepper motor driver to the DIR+ (+5V) terminal. When done, your three stepper motor drivers should look like Figure 6-19.

Figure 6-18. *Wiring the first stepper motor driver for power*

Figure 6-19. *All three stepper motor drivers wired for power*

Wiring Motor Drivers to the Breakout Board

Look closely at the long row of terminals on your breakout board (it should be the longest run of screw terminals). You will see that these terminals are numbered 2, 3, 4, 5, 6, 7, 8, and 9. In between some numbers you should also see COM. We're only concerned at this point with the numbered terminals. For this section, you're going to cut a series of black wires (the length of the wires will vary) and connect them from the breakout board to terminals on the motor drivers. Cut one black wire that will run from the breakout board's 2 terminal to the third motor driver's PUL− terminal. Cut a second black wire that will run from the breakout board's 3 terminal to the third motor driver's DIR− terminal. Figure 6-20 shows these two wires running from the breakout board to the third motor driver.

Figure 6-20. *The third motor driver connected to the breakout board*

Perform the same operation for the first and second motor drivers. For the second motor driver, you will connect the breakout board's 4 terminal to PUL− and the 5 terminal to DIR−. For the first motor driver, you will connect the breakout board's 6 terminal to PUL− and the 7 terminal to DIR−. Figure 6-21 shows all three stepper motor drivers wired to the breakout board.

Figure 6-21. *All three stepper motor drivers connected to the breakout board*

Connecting Power to Motor Drivers

Next, you're going to connect wires from the power supply to the stepper motor drivers. This is the power that will drive the actual stepper motors that will be wired up shortly. Start by cutting a piece of black wire that will run from one of the V+ terminals on the power supply to the VCC+ terminal on the first stepper motor driver. Cut a second piece of black wire and run it from one of the V– terminals on the power supply to the GND– terminal on the first stepper motor driver. This can be seen in Figure 6-22.

Next, cut a small piece of black wire that will be daisy-chained from the VCC+ terminal on the first stepper motor driver to the VCC+ terminal on the second motor driver. Cut another piece of black wire and daisy-chain it from the GND– terminal on the first stepper motor to the GND– terminal on the second stepper motor. Perform the same daisy-chaining operation between the second and third stepper motor drivers—connect their VCC+ and GND– terminals using pieces of black wire. Figure 6-23 shows all three stepper motor drivers with the power supply connected to the first motor driver and daisy-chained wires for connecting the second and third motor drivers.

Figure 6-22. *Connecting the first stepper motor driver to the power supply*

Figure 6-23. *All three stepper motor drivers connected to the power supply*

Connecting Stepper Motors to Motor Drivers

Now it's time to connect the stepper motors to their respective drivers. Earlier in the chapter we told you that it would be helpful to use specific colors for wiring up the wire pairs coming out of each motor. Here's where that will become useful.

Take a close look at a motor driver and you will see four terminals on it, labeled A+, A–, B+, and B–. You will be connecting the four 8' wires you soldered to the motors to these terminals.

CAUTION It is very important that you connect the proper wires from each motor to the correct terminal. Connecting a wire meant for A+ to B+ or B– is bad . . . it can cause immediate damage to a motor when power is supplied. Double-check all your wiring when done with this section.

Take a look at Figure 6-24. We've labeled each of the four wires coming from a motor and also indicated the proper terminal to connect them.

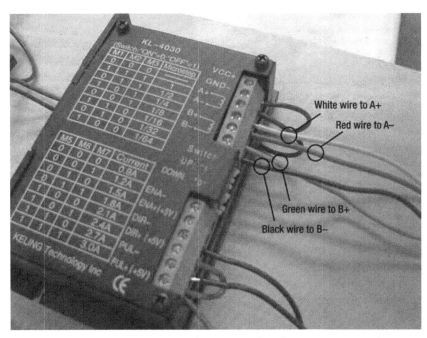

Figure 6-24. *The motor wires must be connected to the correct terminal on a motor driver.*

Your wiring colors may be different, so remember that according to our wire color scheme, the following connections are made:

White 18-gauge wire soldered to the RW/BW pair goes to A+

Red 18-gauge wire soldered to the B/R pair goes to A–

Green 18-gauge wire soldered to the BkW/GW pair goes to B+

Black 18-gauge wire soldered to the Bk/G pair goes to B–

In order to keep our 8' wires from getting tangled, we used plastic conduit (the kind that's cut down the middle so you can insert the wires easily). Figure 6-25 shows one of our three motors connected to the motor drivers and all the wiring safely tucked into the flexible conduit.

Next, you'll need to configure eight small dip switches (toggles indicated in Figure 6-26) on each stepper motor driver. Each switch is numbered, left to right, from 1 to 8.

Figure 6-25. *Conduit will keep the wire from getting tangled.*

Figure 6-26. *Configure the dip switches on each stepper motor driver.*

If a toggle is in the down position, this indicates a value of 0. If the toggle is in the up position, this is a value of 1.

For our CNC machine, we're going to use a microstep setting of 1/4 and a current of 3 amps—each stepper motor has a small diagram printed on its top that gives a 0 or 1 value for each of the eight switches. If you look carefully at Figure 6-24, you can see that a microstep setting of 1/4 corresponds to a value of 101 (for M1, M2, and M3, respectively) and a value of 111 (for M5, M6, and M7, respectively). (Switch 4 and 8 are not used and are set to 0—the down position.)

So, as shown in Figure 6-26, from left to right will be up, down, up, down, up, up, up, down—or 10101110. Be sure to configure all three stepper motor drivers with these toggle switch settings.

Wiring the Cooling Fan

The stepper motor drivers can get quite hot when the machine is in use, so we suggest that you wire up a simple little cooling fan like the kind you would install in a computer case. For ours, we stripped the ends off of a 120V fan, cut the end off of a 5' extension cord (with just two prongs), and then stripped about 3/4" from the ends of the two wires. We then soldered the fan's wires to the two extension cord wires (it doesn't matter which wire is soldered to which); you can see the final result shown in Figure 6-27. The fan will plug into a power strip and can be moved around to get the best air flow to the stepper motor drivers.

Figure 6-27. *The fan will help keep your stepper motor drivers cool.*

Testing the Electronics

The last thing we need to do in this chapter is simply test our wiring. Before you plug in anything, pick up each of your three motors and rotate the shaft by hand. You should find that each motor shaft can be turned easily. Next, go ahead and plug in the 5V (or 6V) adapter—you should see the red LED light on the breakout board light up. If it does not, go back and check your wiring carefully, especially the wiring going into the screw terminal and coming out of it and into the breakout board. You should also see a red LED light up inside each of the motor drivers.

Next, plug in the extension cord that provides power to the power supply. You may or may not hear one or all of the motors "energize." When we plugged our power supply in, each motor "jumped" a fraction of an inch as the motors engaged. Once power is supplied to the motors, you will find that you cannot rotate the motor shafts by hand. Do not try and rotate the motor shafts—this could damage the motor. If all three motor shafts are "locked" and cannot be turned by hand, you've wired them up properly and they are receiving power. Congratulations.

If one or more motors can rotate by hand, power everything off and check your wiring again, especially the wiring done in this chapter's "Connecting Stepper Motors to Motor Drivers" section. Make sure you've connected all the A+, A–, B+, and B– terminals to their proper wire pairs (RW/BW, R/W, BkW/GW, B/G).

Once you've verified all the wiring, power everything off and put all the electronics somewhere safe for now.

TIP For a more detailed wiring diagram, visit `www.buildyourcnc.com/book.aspx` and download the PDF file by clicking the Chapter 6 link.

What's Next?

OK, that's enough wiring for now. We're not quite done with the electronics yet—there's a safety stop switch (also called a *kill switch* or *eStop*, for "emergency stop") that can quickly disable power to your machine that we'll be adding in later in Chapter 19. But now it's time to get cutting, drilling, and assembling the CNC frame.

X-Axis, Part 1

Back in Chapter 1, we introduced you to the general concept of the CNC machine. You learned about the various axes that are used to get the machine's router bit to the material for cutting and drilling. And in Chapter 2, you got a brief overview of the CNC machine you'll be building.

Up to this point, you've been reading and learning about some of the methods you'll be using to cut, join, and build your CNC machine. In this chapter, however, it's time to get your hands dirty (just a little bit, we promise). You're going to begin building the first component for your CNC machine.

You're going to cut some MDF and use the information you learned in Chapters 4 and Chapter 5 to build the x-axis. We don't want to overwhelm you, so we're going to break up the assembly of the x-axis into multiple chapters. Read through this chapter for an overview of what you'll need to accomplish by chapter's end—maybe even reread it. (You're in no rush, right?) Let's get started.

The X-Axis MDF Parts

For your x-axis, you're going to need to cut a total of four pieces of MDF. We will be using a combination of actual part names as well as part letters when describing an MDF part. This will, hopefully, make the instructions a little easier to read (and much less annoying since we won't have to reference each part by its full name).

The MDF parts you will be cutting are as follows:

Part Y	X-Axis Table (top half of table—2'×4')
Part Z	X-Axis Table (bottom half of table—2'×4')
Part T	X-Axis Table Leg
Part U	X-Axis Table Leg

NOTE Please refer to the MDF Parts Layout 1 and MDF Parts Layout 2 PDF files available for download at www.buildyourcnc.com/book.aspx for part names and letters. Refer to the MDF Plans and Cut List PDF file for cutting and drilling dimensions of all MDF parts—this file can also be downloaded at www.buildyourcnc.com/book.aspx.

Again, we highly encourage you to read through this entire chapter first before you begin cutting. Once you're ready to begin, we'll start with the X-Axis Table (Parts Y and Z).

The X-Axis Table

Figure 7-1 shows an image of the final CNC machine's MDF parts all assembled. The table is the largest component of the machine, measuring 24"×48". This surface area consists of two pieces of MDF, both 24"×48", bolted together to provide a strong surface and reduce warping. (The tabletop also has angled rails bolted to the two long sides of the MDF—this will be covered in Chapter 8.)

For this chapter, we're going to focus on only the tabletop; the table's legs (Parts T and U) will be covered in Chapter 8.

CAUTION The part layouts found in the MDF Parts Layout 1 and MDF Parts Layout 2 PDFs *are not to scale*; refer to the MDF Plans and Cut List file for actual dimensions.

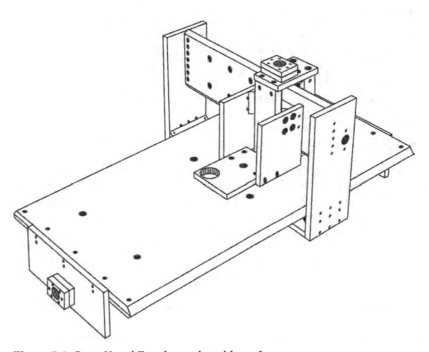

Figure 7-1. *Parts Y and Z make up the table surface.*

If you've purchased a large 48"×96" piece of MDF, you'll need to cut that sheet into four equal parts each measuring 24"×48". If your hardware/lumber supplier provides MDF in quarter sheets (24"×48"), like the ones shown in Figure 7-2, this will reduce the amount of cutting you'll have to perform. The cost of four quarter-sheets may be a little higher than a single 48"×96" sheet, but you'll save some time in not having to measure and cut a larger sheet.

NOTE Because the saw blade has a thickness (typically 1/8"), anytime you cut your MDF you'll be losing a small amount of material. If you mark a line in pencil and cut down that line as accurately as possible, the piece you are cutting will be shy (short) about 1/16". Try to maintain a consistent method when cutting—will you try to cut exactly down the center of the line, to the left, or to the right? Whatever your decision, use it with every cut you make. Overall, this design will allow for small deviations caused by the thickness of the cutting blade.

Figure 7-2. *Quarter-sheets of MDF will save some cutting time—two of these make the table.*

NOTE At our hardware store, we were able to purchase quarter-sheets of MDF for $7.95 each. A larger sheet of MDF with dimensions of 49"×97" was sold for $25.98. Besides the price difference of around $6.00, the larger sheet was also oversized a little beyond the 48"×96" mentioned earlier. Sometimes the edges of the larger sheets of MDF are a little dented or broken, so the larger MDF sheets come slightly oversized so you can cut away the damaged portions. Keep this in mind when cutting the large MDF sheet into four parts—each of those four smaller sheets needs to be 24"×48".

Once you have two sheets of MDF with dimensions of 24"×48", consult the MDF Plans and Cut List PDF file for help in laying out the pieces for drilling. You'll be referencing the pages labeled "Bottom Half of Table" and "Top Half of Table" for the measurements used in drilling and chamfering the two pieces. (If you're not familiar with chamfering yet, we'll explain that shortly.)

Figure 7-3 shows a portion of the Bottom Half of Table sheet and Figure 7-4 shows a portion of the Top Half of Table sheet.

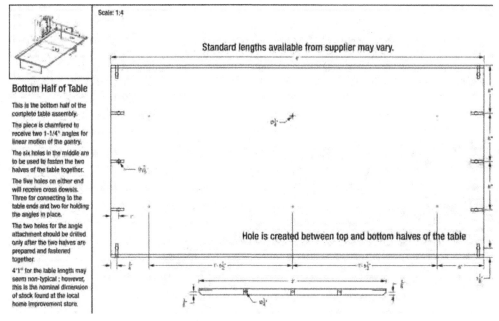

Scale: 1:4

Standard lengths available from supplier may vary.

Bottom Half of Table

This is the bottom half of the complete table assembly.

The piece is chamfered to receive two 1-1/4" angles for linear motion of the gantry.

The six holes in the middle are to be used to fasten the two halves of the table together.

The five holes on either end will receive cross dowels. Three for connecting to the table ends and two for holding the angles in place.

The two holes for the angle attachment should be drilled only after the two halves are prepared and fastened together.

4'1" for the table length may seem non-typical ; however, this is the nominal dimension of stock found at the local home improvement store.

Hole is created between top and bottom halves of the table

Figure 7-3. *The bottom half of the table.*

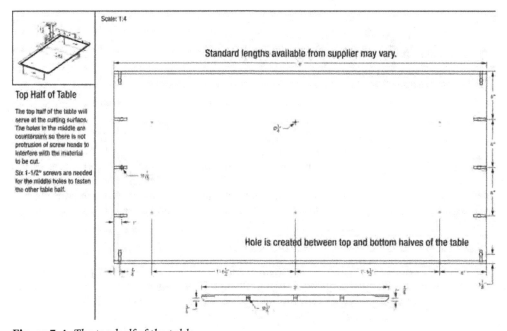

Scale: 1:4

Standard lengths available from supplier may vary.

Top Half of Table

The top half of the table will serve at the cutting surface. The holes in the middle are countersunk so there is not protrusion of screw heads to interfere with the material to be cut.

Six 1-1/2" screws are needed for the middle holes to fasten the other table half.

Hole is created between top and bottom halves of the table

Figure 7-4. *The top half of the table.*

76

Because you will be bolting the top and bottom halves of the table together, it will be easier work if you clamp the two pieces together, mark the spots to be drilled on the top sheet of MDF, and then drill pilot holes (1/8") through the top and bottom sheets at the same time. Both the top and bottom halves will have larger holes drilled in identical places, so by clamping the two pieces together and drilling pilot holes, you can be assured that when you bolt the two pieces together, all the holes will match up.

After you've marked the top half of the table and drilled all the pilot holes, it's time to chamfer the edges. What is a chamfered edge? Look at Figure 7-5.

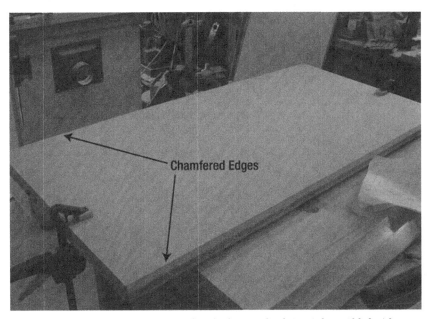

Figure 7-5. *The tabletop has chamfered edges on both its right and left sides.*

The chamfering is done using your router and a special bit (called a chamfer bit). It allows you to cut away material at an angle—in this case, a 45 degree angle. Refer back to Figures 7-3 and 7-4 and you can see a profile of each piece. Viewed from the side, there is a small 45 degree surface. This angled surface will allow you to attach a piece of angled aluminum rail to the sides of the table; you'll learn more about this in Chapter 8.

CAUTION Use a router with extreme caution. Consult your router's documentation for proper usage and safety. Use safety goggles and remember to always move the router in the same direction that the bit spins.

Chamfer the top edges on the top half as indicated by the CNC plans, and chamfer the bottom edges of the bottom half, also as indicated in the CNC plans. (This entails using the 45 degree chamfer bit to remove half of the 3/4" sides of Parts Y and Z. It doesn't have to be exact, but try to be as accurate in your chamfering as possible.)

If these two pieces were clamped together now, the side of the tabletop would look like Figure 7-6.

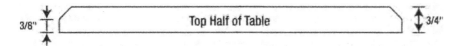

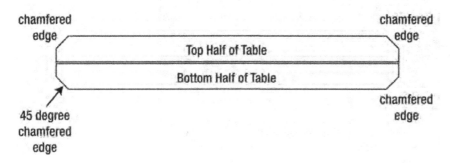

Figure 7-6. *The tabletop with chamfered edges as seen from the end*

Once you've chamfered the edges, it's time to make the table's rails.

Cutting Rails for Tabletop Sides

The last task for this chapter is the cutting of two pieces of angled aluminum rail. You can use either 1" or 1 1/4" rail for the sides; for our tables, we selected 1 1/4" rail and cut two pieces, each 48" in length. Figure 7-7 shows one piece of rail cut.

TIP If you use 1" rails for the x-axis tabletop, this is the only time you'll be able to use the 1" angled aluminum rail; when the rail is mounted to the tabletop, its edges won't protrude above or below the tabletop surface, allowing you to use a 90 degree square later when mounting the gantry sides (see Chapters 10 through 12). We didn't have any 1" rails at the time, so we went with 1 1/4", which also works fine but does create a raised lip just slightly above the tabletop's surface.

Take the two 48" pieces of rail and set them aside for now.

Figure 7-7. *Angled aluminum rail cut to fit the sides of the table*

Summary of Work

At this point, you should have the following items completed:

> Part Y with dimensions of 24"×48"
>
> Part Y chamfered along the top of the 48" edges
>
> Part Z with dimensions of 24"×48"
>
> Part Z chamfered along the bottom of the 48" edges
>
> Two 1 1/4" rails cut to 48" length (1" rail can be substituted)
>
> Parts Y and Z marked and 1/8" pilot holes drilled

What's Next?

In Chapter 8, you're going to continue work on the table by drilling holes in the MDF to bolt the top and bottom halves of the table together. Once the two halves are joined, you'll drill additional holes that will be used in Chapter 9 to attach the table's legs and the rail.

X-Axis, Part 2

In this chapter, we'll continue working on constructing the largest part of your CNC machine—the table. You're going to drill a variety of holes in the two halves that make up the tabletop, including holes for cross dowels that will allow you to later connect the legs to the table.

Read through the entire chapter first so you'll understand what work you'll complete next. By the end of this chapter, you will have bolted together the two tabletop halves and cut the table legs. Chapter 9 will then show you how to add the legs and the angled aluminum rail on the sides of the table. Let's get started.

Drilling the Table

At this point, you should have Parts Y and Z clamped with pilot holes drilled using a 1/8" bit. All of these holes will need to be drilled with larger drill bits, but there is a mixture of sizes and types of drilling to be done, so let's go slowly here.

Take a look at Figure 8-1. This shows the surface of Part Z. The six holes indicated are for bolts that will be inserted into Part Y and through Part Z, and secured using nuts on the bottom of Part Z. These six bolts will hold Parts Y and Z together securely. The holes you will drill here must be countersunk using a 1/4" countersink drill bit. This will allow the bolts to sit just below the flat surface of the table.

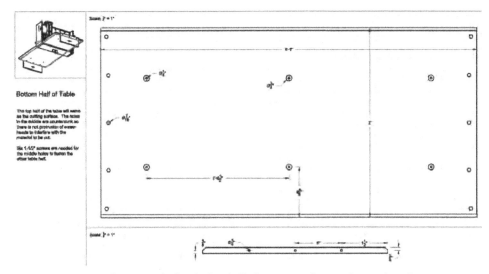

Figure 8-1. *Part Z has a total of 16 holes drilled on its surface and 6 on the edges.*

In Figure 8-1, the ten 7/16" holes drilled (five on the left edge and five on the right edge) are for inserting cross dowels. Insert a 7/16" drill bit and drill those ten holes using the pilot holes you drilled in Chapter 7. You'll also drill three 1/4" holes on the left edge and three 1/4" holes on the right edge where bolts will be inserted through the table legs and screwed into the cross dowels.

Figure 8-2 shows how Part Y will look when the 1/4"-20 X 1" countersink Phillips flat head screws, or tapered bolts, are inserted—we're not yet ready for this step; this photo is just so you can see how the countersunk holes will allow the 1" bolt heads to sit just below the surface. (You can also see the holes drilled on the surface for the cross dowels and the three holes drilled into the edges of Parts Y and Z where the table legs will attach—more on that later.)

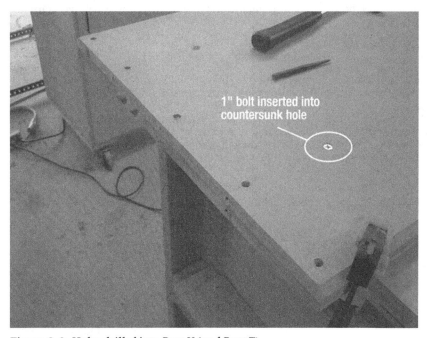

Figure 8-2. *Holes drilled into Part Y (and Part Z)*

Set Part Y aside for a moment and grab Part Z. You'll be drilling holes in the bottom of Part Z (the surface that will be facing downward once the two halves are attached together). Using a 7/16" drill bit, drill the ten cross dowel holes as indicated in the plans (five on the left and five on the right). Drill 1/4" holes in the six holes that will be used to bolt the top and bottom table halves together. You can see Part Z in Figure 8-3 with these holes drilled.

Figure 8-3. *Part Z with cross dowel holes and 1/4" bolt holes drilled*

Next, you're going to need to counterbore the six 1/4" holes to a depth of around 1/4" and a diameter of 3/4"—this is shown in Figure 8-4. We recommend a 3/4" Forstner bit for this task, and it may require testing on some scrap pieces of wood to get the right depth set. You'll want to drill it deep enough so that the nut inserted into the counterbore will sit just below the surface.

NOTE You do not have to counterbore these holes, but you will have to insert a longer bolt (1 1/4") to connect Parts Y and Z. Just be certain that the length of bolt you use will not interfere with any moving parts on the final CNC table.

Figure 8-5 shows how the bottom of Part Z will look when the nuts are added; notice they sit just below the surface.

Figure 8-4. *Counterbore the six holes on the bottom of Part Z.*

Figure 8-5. *The bottom of Part Z with 1/4" nuts inserted into counterbored holes*

Next, attach the two halves together. This is done by inserting six 1" tapered bolts (1/4" diameter) into the six holes on the surface of Part Y, and using 1/4" nuts to secure them. Tighten the nuts securely but don't overtighten.

TIP If you find that one or more of your holes do not line up exactly, you can enlarge the 1/4" holes by drilling them out with a 5/16" drill bit if necessary.

Drilling Holes for Legs

Now that Parts Y and Z are attached, it's time to drill the holes for the table legs. Take a look at Figure 8-6. We've used a straight edge to extend lines from the five holes on the surface of Part Y to the table's edge. This is done so we can drill these holes to match up with the 7/16" holes that will hold the cross dowels. We've also drawn two straight horizontal lines halfway through the edges of both Parts Y and Z.

Figure 8-6. *The tabletop with chamfered edges as seen from the end*

Figure 8-7 shows a close up of the holes drilled on the table's edge. There will be six on one side and six on the other (three per half). Notice the lines drawn to help match up the proper location for drilling these holes.

Figure 8-7. *Holes drilled into the table's edges will let you attach the legs.*

Cutting the Table Ends

Before we finish this chapter, we'll end by cutting out two ends (legs) for the table, Parts T and U. After cutting the legs, mark them as instructed in the CNC plans. Figure 8-8 shows the plans for cutting and drilling the legs.

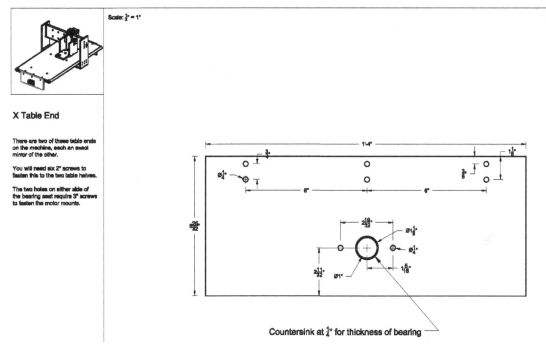

X Table End

There are two of these table ends on the machine, each an exact mirror of the other.

You will need six 2" screws to fasten this to the two table halves.

The two holes on either side of the bearing seat require 3" screws to fasten the motor mounts.

Countersink at $\frac{1}{4}$" for thickness of bearing

Figure 8-8. *Marking the table legs for drilling*

Figure 8-9 shows one of the legs with drilling marks partially completed. Notice that we haven't yet drawn the marks for the larger hole on Parts T and U.

Figure 8-9. *After marking the legs, double-check the measurements, as accuracy is needed here.*

The two legs must be centered as perfectly as possible for the CNC machine to work properly. To help with this, we recommend extending the pencil marks shown in Figure 8-9 to the top edge of Parts T and U. Check to make sure that these pencil marks line up with the 1/4" holes on the table edges, as shown in Figure 8-10.

If the holes used to mount the legs to the table line up, finish up by marking the legs for the larger center hole to be drilled as shown in Figure 8-11. Extend the line for the 1/4" bolt used to connect the leg to the table down. This will ensure that the larger hole (see Figure 8-8 again) on Part T lines up exactly with Part U after the legs are connected to the table.

Figure 8-10. *Make certain the legs are centered properly.*

Figure 8-11. *All the drilling marks completed for the table legs*

89

Summary of Work

At this point, you should have the following items completed:

    Parts Y and Z bolted together using 1/4" tapered bolts

    Parts Y and Z drilled with cross dowel holes and holes on edges for attaching legs

    Parts T and U cut to proper size

    Parts T and U marked for drilling

Hardware Required

For the work performed in this chapter, you will use

    1/4" tapered bolts; 1" length; quantity: 6

    1/4" nuts; quantity: 6

What's Next?

In Chapter 9, we'll finish building the CNC tabletop by drilling the legs, attaching them, and then adding the angled aluminum rail to the sides. When done, you'll have approximately one-third of your CNC machine completed!

X-Axis, Part 3

You'll reach an important milestone in this chapter—the completion of the CNC machine table. The table is the largest portion of your CNC machine, and if you've followed the instructions in Chapters 7 and 8, you're ready to finish up the work required to assemble it.

Later chapters in the book will have you cutting and drilling additional MDF pieces that will be added to the table. Keep in mind that the table is 2'×4' in size, so you'll want to put it somewhere in your work area that is easily accessible (preferably above waist level to make it easier to work on) while still allowing you some free room to measure, cut, and drill. You can certainly use the table surface for this kind of work, but just be careful not to drill or cut into the table surface or damage it by accidentally hitting it with other MDF parts.

We'll start by finishing up the drilling of the table's legs and then attach these legs to the table. When done, you can congratulate yourself on the successful completion of one-third of your final CNC machine.

Drilling the Table Ends (Legs)

You should have Parts T and U marked for drilling after completing Chapter 8. Before you begin drilling any holes, we'd like to draw your attention to the larger hole in the center of both pieces, as shown in Figure 9-1.

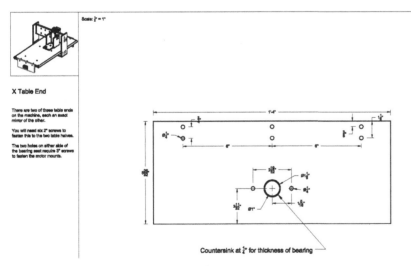

Figure 9-1. *The larger hole in the center must be drilled carefully.*

NOTE Refer to the MDF Plans and Cut List PDF file for cutting and drilling dimensions of all MDF parts—this file can be downloaded at www.buildyourcnc.com/book.aspx.

Don't worry about the drilling yet. Just take a look at Figure 9-2, which shows a closeup image of that larger center hole drilled. Notice that it consists of a smaller-diameter hole drilled completely through the piece of MDF and a larger-diameter hole drilled approximately halfway through the piece. This creates a *shoulder*, or *shelf*, that will be used to support a bearing that is inserted later in the building process. We'll explain how to drill this hole in more detail shortly.

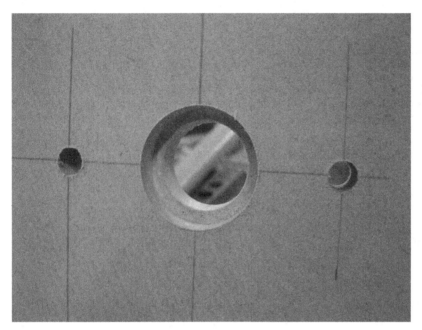

Figure 9-2. *The large center hole consists of a smaller-diameter hole within a larger-diameter hole.*

For now, go ahead and drill the 1/4" holes in Parts U and T, as shown in Figure 9-3. (For best results, drill pilot holes using a 1/8" bit in all holes, including the larger center hole, before drilling the larger holes.)

These 1/4" holes are where you will insert 2" bolts to secure the legs to the tabletop. (These bolts will screw into cross dowels inserted into the holes drilled on the tabletop.)

Next, you're going to need to counterbore the larger hole first. The diameter of the larger hole is 1 1/8". Drill to approximately 3/8" depth. The easiest method for doing this is to use a Forstner bit. Figure 9-4 shows Part U with the larger-diameter hole drilled.

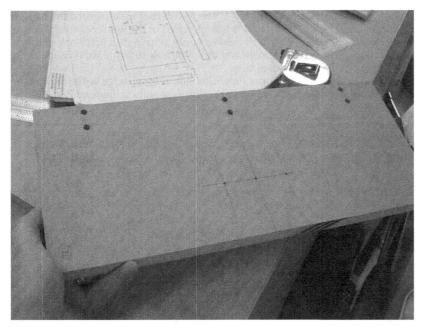

Figure 9-3. *Part U with the 1/4" holes drilled for connecting to the tabletop*

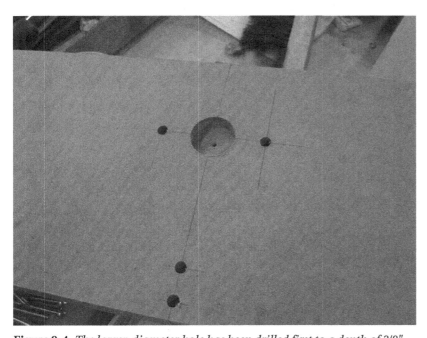

Figure 9-4. *The larger-diameter hole has been drilled first to a depth of 3/8".*

There are two things we'd like to point out about Figure 9-4. The first is that we've drilled the two 1/4" holes to the right and left of the larger hole. The second is the "dimple" in the center of the larger hole. This hole was created by drilling a pilot hole. The pilot hole not only helps a drill bit to self-center, but will also be useful to us when drilling the inner hole; without the pilot hole to guide us, it would be difficult to drill the inner hole centered in the larger outer hole.

The plans call for using a 1"-diameter bit to drill the inner hole. This will create an edge inside the larger hole of only 1/16" width. Figure 9-5 shows the final results of Part T. Perform the same actions on Part U.

TIP While the 1/16" shoulder is sufficiently wide to support a bearing that will be inserted into the hole later, we used a 3/4" Forstner bit to create a slightly smaller inner hole but a larger shoulder (edge). Instead of a 1/16" shoulder inside the larger hole, we now have a 3/16" shoulder for the bearing to rest on.

Figure 9-5. *Part U with all holes drilled, including the larger center hole with shoulder*

Now, before you attach Parts T and U, you need to attach the 4' rail you cut in Chapter 8 that will attach to the sides of the table (Parts Y and Z). Go ahead and set Parts T and U aside for a moment and grab the two 4' lengths of aluminum angled rail.

Drilling and Mounting the Rail

Mark both pieces of 4' rail 3/4" from the ends. Drill 1/4" holes into the rails from the inside out, not the outside in. (You may find that drilling the holes using a slightly larger bit—5/16" is a good size—will allow for a little wiggle room for the inserted bolts to find the holes on the sides of the tabletop.)

TIP You may wish to build a jig to hold the rail while drilling. One way to do this is to clamp two small pieces of waste MDF to the table (or drill press table) with a small 1/2" gap between the pieces. Place the rail in the gap so the inside edges are facing up for drilling. You can also secure the rail with another clamp to one or both of the scrap MDF pieces to keep it from moving.

Figure 9-6 shows a piece of 4' rail with one hole drilled 3/4" from the end.

Figure 9-6. *The tabletop with chamfered edges as seen from the end*

After drilling two holes into a piece of rail, fit it into place over the chamfered edge of the tabletop and mark holes for drilling. Place a clamp on Parts Y and Z to hold the pieces tightly together, drill 1/4" holes into the sides of the chamfered edges, and use 2" bolts to bolt them to the tabletop using cross dowels inserted into the 7/16" holes on the tabletop surface.

Figure 9-7 shows the two pieces of 4' rail attached to the sides of the table. (A cross dowel is located inside the hole indicated by the arrow, but is just below the surface and not visible.)

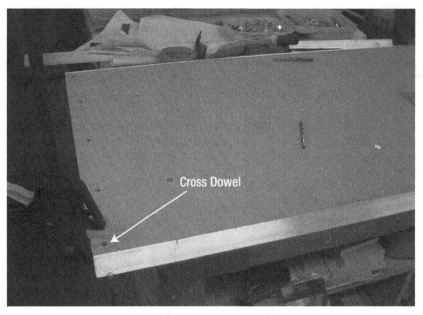

Figure 9-7. *Two rails attached to the sides of the table*

After the rails are attached, it's time to attach the legs to the table.

Attaching the Table Legs

All that's left is to connect one leg to each end of the table. Use six 2" bolts with six matching cross dowels per leg. Three bolts and three cross dowels will be used on the top and three bolts and three cross dowels will be used on the bottom. Figure 9-8 shows all six bolts connecting one leg to the table with the three cross dowels on top visible. (You may find it easier to screw in the bottom three bolts to the cross dowels by turning the table on its side.)

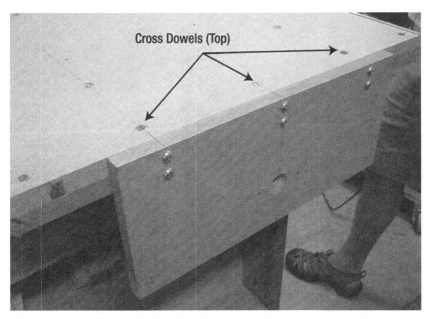

Figure 9-8. *One leg attached to the table using bolts and cross dowels*

Cutting the X-Axis Lead Screw

When your machine is done, it will use three different lengths of threaded rod to assist with the movement of the router—these are called *lead screws*. One piece of lead screw will be used for each axis, so you'll want to go ahead and purchase enough threaded rod to cut all three pieces. We recommend that you purchase one 8' length of 1/2"-diameter regular threaded rod with 13 threads per inch (TPI). If you can find a hardware store willing to cut them to the lengths required, consider yourself lucky; usually you'll need to purchase them in precut lengths and then cut them to your desired length later.

TIP For our machine, we were able to use a single 8' rod to cut all three pieces. If you have to purchase the rod in shorter lengths, remember that the x-axis lead screw will be longer than 4', so try and purchase your lead screws in 6' or 8' lengths. The safest purchase will be to purchase a single 6' length and a 3' length if possible. The 6' length will be enough for the x-axis and z-axis lead screws, and the 3' length will be enough for the y-axis lead screw.

The length of the x-axis tabletop is 4' 1.5" (4 feet, 1.5 inches)—the lead screw will be inserted through two bearings inserted into the large holes drilled in Parts T and U. Mark and cut an extra 1" of lead screw protruding from both ends, as shown in Figure 9-9.

After cutting the x-axis lead screw to length (approximately 4' 3.5"), remove it from the machine and set it aside for later.

Figure 9-9. *Mark an extra 1" of lead screw on both ends of the table and cut.*

Summary of Work

At this point, you should have the following items completed:

> Rails attached to sides of tabletop
>
> Parts T and U drilled with proper holes
>
> Parts T and U mounted to the table
>
> Lead screw for the x-axis cut to a length of approximately 4' 3.5"

Hardware Required

For the work performed in this chapter, you will use:

> 1/4" pan head bolts; 2" length; quantity: 16
>
> Cross dowels; quantity: 16
>
> Lead screw option 1: 1/2"-diameter regular threaded rod; 13 TPI; 8'
>
> Lead screw option 2: two 1/2"-diameter regular threaded rods; 13 TPI; 6' and 3'

What's Next?

You've completed the base of your CNC machine. All other work done in later chapters will build onto the table. Your x-axis is completed, so you're now ready to move on to the y-axis. Chapter 10 will get you started on a few of the parts needed to give your CNC machine the ability to move its mounted router front to back along the 4' rails you just mounted.

Y-Axis, Part 1

Now that you have the largest part of your CNC machine built—the table—you've completed what we refer to as the x-axis. In order for the router to do its work, you've got two more axes to complete: the y-axis and the z-axis.

What we're going to show you in this chapter are the first steps to building the y-axis and giving your CNC machine's router the ability to move forward and backward, toward the front and back of the tabletop. To do this, you're going to create a frame around the tabletop that consists of four pieces of MDF and two more bearing-rail assemblies (BRAs).

Over the next three chapters, you're going to build the frame that will give the router movement along the y-axis.

The Y-Axis MDF Parts

For the y-axis (the discussion of which spans Chapters 10 to 12), you're going to need to cut a total of six pieces of MDF. Once again, we'll be using a combination of actual part names and part letters when describing an MDF part. The MDF parts you will be cutting are

Part O   Y-Axis Rail Support

Part P   Y-Axis Gantry Bottom Support

Part Q   Y-Axis Gantry Side

Part R   Y-Axis Gantry Side

Part S   Y-Axis Rail Reinforcement

Part E   Y-Axis Gantry Bottom Nut

NOTE Please refer to the MDF Parts Layout 1 and MDF Parts Layout 2 PDF files available for download at www.buildyourcnc.com/book.aspx for part names and letters. Refer to the MDF Plans and Cut List PDF file for cutting and drilling dimensions of all MDF parts—this file can also be downloaded at www.buildyourcnc.com/book.aspx.

Please read through this entire chapter before you begin cutting. Once you're ready to begin, we'll start with the Y-Axis Gantry Sides (Parts Q and R).

Parts Q and R: The Y-Axis Gantry Sides

Figure 10-1 shows an image of the final CNC machine's MDF parts all assembled with the Y-Axis Gantry Sides circled.

For this chapter, we're going to focus on cutting and drilling these two pieces.

CAUTION The part layouts found in MDF Parts Layout 1 and MDF Parts Layout 2 *are not to scale*; refer to the MDF Plans and Cut List file for actual dimensions.

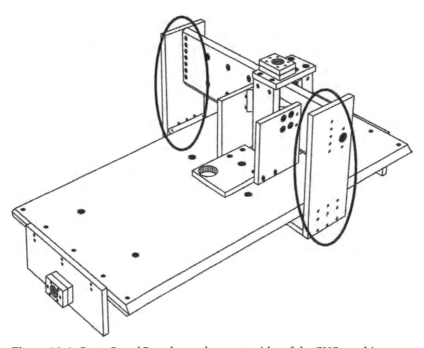

Figure 10-1. *Parts Q and R make up the gantry sides of the CNC machine.*

Figure 10-2 shows Part R cut to the proper dimensions of 17 3/4"×7". Part Q is cut to the same dimensions, and both pieces have been marked for drilling. Refer to the CNC plans PDF file for the locations and measurements of the holes to be drilled.

We've mentioned this earlier in the book, but when drilling holes, it is sometimes helpful to use a center punch, as shown in Figure 10-3. You can create a small dimple in the MDF to help you center the drill bit with more precision.

Figure 10-2. *Part R cut and marked for drilling*

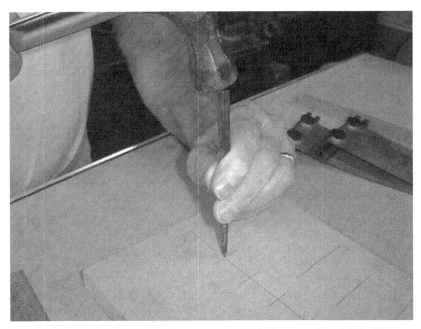

Figure 10-3. *Use a center punch to create small holes to assist with drilling.*

Next, clamp parts Q and R together, as shown in Figure 10-4. You can drill the two parts separately, but clamping them together saves time and can help to ensure that when the two parts are mounted, their respective holes will match up.

Figure 10-4. *Clamp parts Q and R together before drilling.*

You're going to drill a total of 16 1/4" holes in the gantry sides before they are mounted. (The larger holes shown in Figure 10-1 will hold bearings and will be drilled into each gantry side shortly.) Figure 10-5 shows all 16 1/4" holes drilled into Parts Q and R.

TIP We used a 3/32" drill bit to drill pilot holes prior to drilling the larger 1/4" holes.

Next, you're going to drill the larger hole into each gantry side that will hold a bearing. Refer back to Chapter 9 for instructions on drilling the shoulder and large hole that will hold a bearing. The holes for the bearings need to be drilled as accurately as possible. For that reason, here is an alternative method for drilling these holes with a drill press that differs slightly from the method used in Chapter 9.

First, keep the two pieces clamped together as shown in Figure 10-5. Use a 1" Forstner bit to drill a single hole through both pieces, as shown in Figure 10-6.

Figure 10-5. *All of the 1/4" holes have been drilled in Parts Q and R.*

Figure 10-6. *Drill a 1" hole through Parts Q and R.*

Next, while keeping the two pieces clamped together, counterbore the larger hole on Part Q, as shown in Figure 10-7. The diameter of the larger hole is 1 1/8". Drill to approximately 3/8" depth. Once again, the easiest method for doing this is to use a Forstner bit.

Figure 10-7. *Drill a 1 1/8" hole into Part Q.*

After drilling the larger hole into Part Q, flip the clamped pieces over. Put the 1" Forstner bit back in the drill (or drill press) and, without turning the drill press on, lower the 1" bit until it goes into the 1" hole in Part R. Clamp Parts Q and R to your drill press table. This will allow you to remove the 1" Forstner bit, insert the 1 1/8" Forstner bit, and drill the second larger hole centered perfectly in part R.

Figure 10-8 shows Parts Q and R with the bearings inserted into the holes. A 1/2" rod has been inserted through the bearings to show that the holes match up.

Now set Parts Q and R aside for a moment, and let's build two BRAs, one for each gantry side.

Figure 10-8. *A 1/2" rod inserted to check that the 1 1/8" holes are lined up properly*

Building BRAs for Gantry Sides

Each gantry side will have a BRA mounted to it. Refer to Chapter 4 for details on building BRAs; for this chapter, you're going to cut two 7" lengths of angled aluminum rail. Figure 10-9 shows the two pieces cut.

Figure 10-9. *Cut two 7" lengths of angled rail for the BRAs.*

Next, you'll need to use your pilot hole template (see Chapter 4) to drill 1/8" pilot holes at the ends of each piece of rail. Figure 10-10 shows the pilot holes being drilled into the rails using the small metal template.

Figure 10-10. *Drilling pilot holes into the BRA rails*

Next, use the 17/64" bit to drill holes for tapping. Figure 10-11 shows the holes drilled and ready to be tapped.

Figure 10-11. *The larger holes are drilled into the rails and ready to be tapped.*

Finally, use the tap to add the threads to the two pieces of rail. Figure 10-12 shows the tap inserted into one of the pieces of rail. Go slowly, and make sure that the tap bit is "biting" into the hole as straightly as possible.

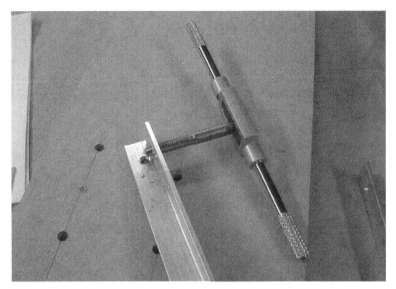

Figure 10-12. *Use the tap to add threads to the holes.*

Figure 10-13 shows the bolt-bearing-nut assemblies screwed into the rails. You now have two 7" BRAs, ready to be attached to the gantry sides.

Figure 10-13. *Add bolt-bearing-nut assemblies to the rails to complete your BRAs.*

Summary of Work

At this point, you should have the following items completed:

Part Q with 16 1/4" holes drilled

Part R with 16 1/4" holes drilled

Part Q with a lead screw bearing hole drilled

Part R with lead screw bearing hole drilled

Two 7" BRAs assembled

Hardware Required

For the work performed in this chapter, you will use

Bearings; 1/2" inner diameter, 1 1/8" outer diameter; quantity: 2 (for lead screw)

Bearings; 5/16" inner diameter, 7/8" outer diameter, 1/4" thick; quantity: 8 (for BRAs)

1/4" bolts; 1" length; quantity: 8 (for BRAs)

1/4" nuts; quantity: 16 (for BRAs)

What's Next?

You'll continue the cutting and drilling of parts for the y-axis in Chapter 11. The BRAs will get mounted so you can attach the sides and take some special measurements for two more parts that make up the y-axis frame.

Y-Axis, Part 2

You've completed cutting and drilling the gantry sides, so now it's time to work on the MDF pieces that will allow you to complete the y-axis frame that will wrap around the tabletop. This will consist of cutting a piece that will be mounted under the tabletop (with its surface parallel to the surface the final CNC machine will sit on) and two pieces that will be mounted above the tabletop.

Chapter 11 will focus on cutting the MDF component that will mount underneath the tabletop. After cutting, drilling, and connecting this piece to the gantry sides, you'll take a special measurement that will help you with the final pieces of the y-axis frame in Chapter 12.

The Y-Axis MDF Parts

As mentioned in Chapter 11, the y-axis that you build in Chapters 10 to Chapter 12 will consist of six pieces of MDF. Those MDF parts are

Part O	Y-Axis Rail Support
Part P	Y-Axis Gantry Bottom Support
Part Q	Y-Axis Gantry Side (completed in Chapter 10)
Part R	Y-Axis Gantry Side (completed in Chapter 10)
Part S	Y-Axis Rail Reinforcement
Part E	Y-Axis Gantry Bottom Nut

This chapter will focus on cutting, drilling, and mounting Part P, the Gantry Bottom Support. The completed CNC machine image shown in Figure 11-1 hides Part P, but we've provided a shaded outline of where the piece will be placed.

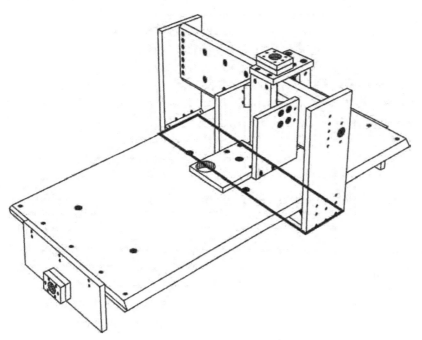

Figure 11-1. *Part P is mounted underneath the table.*

NOTE Please refer to the MDF Parts Layout 1 and MDF Parts Layout 2 PDF files available for download at www.buildyourcnc.com/book.aspx for part names and letters. Refer to the MDF Plans and Cut List PDF file for cutting and drilling dimensions of all MDF parts—this file can also be downloaded at www.buildyourcnc.com/book.aspx.

Please read through this entire chapter first before you begin cutting and drilling. Part P is a critical piece to the smooth operation of your final CNC machine, so take your time and proceed carefully.

Attaching BRAs and Gantry Sides

In Chapter 10, you built two BRAs, each with a length of 7". Those two BRAs need to be attached to the gantry sides at this time. You'll use six 1 1/4" bolts and nuts to secure each BRA to a gantry side. Mount the BRAs so that they ride the 4' rails mounted to the sides of the table. Take a look at Figure 11-2; it shows Part R and the six predrilled 1/4" holes that will hold down the BRA.

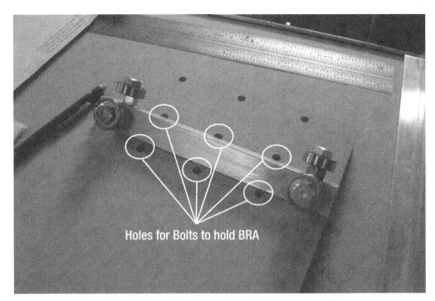

Figure 11-2. *Part R with the six holes used to bolt down the BRA*

Figure 11-3 shows a gantry side with its BRA mounted and able to move along the 4' rail bolted to the table's side.

Figure 11-3. *A single gantry side with a BRA rolling along the table's rail*

For best results, take off the bolt-bearing-nut assemblies before bolting down the 7" rails. Don't tighten down the bolts until you've measured and confirmed that each end of a rail is the same distance from the bottom edge of the gantry side. Figure 11-4 shows the proper locations to measure to ensure that the BRAs are mounted perfectly parallel to the table surface.

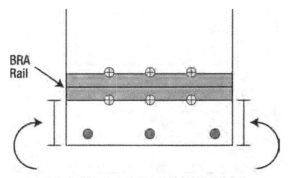

BRA
Rail

Measure distance from Rail bottom edge to
Gantry bottom edge on both sides. Distance
should be the same.

Figure 11-4. *Mount the BRA rails so that they are parallel to the table surface.*

Now, this is where it can get a little tricky—having a second (or third) set of hands will really help. You're going to need to hold the two gantry sides (with BRAs attached) to the table as perfectly vertical as possible, and measure the distance between them under the table. Figure 11-5 shows the two gantry sides being held to the table with clamps, but feel free to use any method you like as long as you can keep the two sides as vertical as possible and not angling inward our outward.

NOTE Notice anything unusual about Figure 11-5? The lead screw holes we drilled in Chapter 10 to hold the bearings aren't done yet! Don't let this bother you—we were building two machines at once. You may see an occasional photo in the book with a step or two missing or out of order. Sometimes this was intentional (to get a better photograph) and sometimes it wasn't. We used our building and testing time to determine the best order for doing all the cutting and drilling so you wouldn't have to! So, don't let the lack of bearings in the lead screw in Figure 11-5 bother you—just keep pushing forward.

Once Parts Q and R are in place, carefully measure the distance between them from under the table. Figure 11-6 shows what you need to measure.

Figure 11-5. *Mount the gantry sides (Parts Q and R) to the table.*

Measure this
Distance

Figure 11-6. *Measure the distance between Parts Q and R.*

Write down the measurement of the distance between Parts Q and R so you don't forget it; you'll need it to cut Part P to the proper length, as well as another MDF piece you'll cut in Chapter 12. If you like, go ahead and write this critical measurement below, on this very page, so that you'll have it when you need it:

Inside distance between Parts Q and R: _____

Part P: The Y-Axis Gantry Bottom Support

If you consult the CNC machine plans, you'll note that Part P has dimensions of 7"×2' 2 11/16" (2 feet, 2 and 11/16 inches). While the 7" height is correct, you're going to use the measurement you just made for the inside distance between Parts Q and R. Cut that single piece of MDF now. Figure 11-7 shows our Part P cut (our measurement was 26 3/4") and marked for drilling.

Figure 11-7. *Part P cut and marked for drilling*

You're going to drill six 7/16" holes for the cross dowels that will be inserted into the surface of Part P—three on the left and three on the right. You'll also drill six 1/4" holes for the 3" bolts that will be inserted into the sides of Parts Q and R and screw into the cross dowels—three on the left edge and three on the right edge. Figure 11-8 shows Part P with the drilling completed.

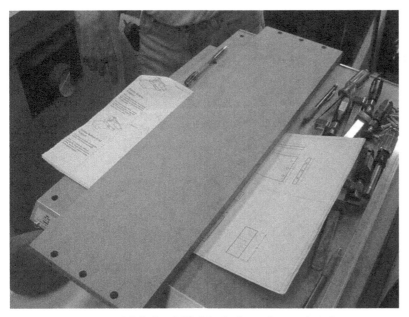

Figure 11-8. *Part P with holes drilled for bolts and cross dowels*

Next, use the plans to measure and mark the holes for drilling Part E. Part E prior to drilling is shown in Figure 11-9.

Figure 11-9. *Part E needs to be drilled and bolted to Part P.*

Next, mark and drill Part P where Part E will be bolted to it. Countersink the holes on Part P as shown in Figure 11-10.

Figure 11-10. *Parts E and P drilled and ready to be bolted together*

Two 3" bolts will be inserted into Part P through Part E, and nuts will secure them, as shown in Figure 11-11.

Figure 11-11. *Parts E and P bolted together*

It's easier to attach Part P (with Part E bolted to it) to Parts Q and R if you take the gantry sides off the table. Don't tighten down on the bolts yet, though—leave them loose so you'll be able to fit the BRAs back over the rails and then tighten them down.

Lift up the table on one end and slide the partial frame under the table leg. Carefully fit the gantry sides so that the BRAs clamp onto the 4' rails. Once the BRAs are fitted in place, tighten down on the bolts on the sides of Parts Q and R to hold the partial frame snuggly to the table. Figure 11-12 shows the partial frame mounted to the table.

Figure 11-12. *Partial frame with gantry sides and bottom installed*

Next, slide the gantry sides and bottom along the rails as far forward as possible, toward one of the table legs. Use clamps to keep Parts P, Q, and R from moving back and forth along the tabletop. Using a hand drill with a 1/2" drill bit, insert the bit into the x-axis leg's lead screw bearing, and drill a horizontal hole through Part E, as shown in Figure 11-13.

TIP Another solution is to use the drill bit to make only a dimple in Part E. Remove Part E and then drill the hole with a drill press.

Remove Part E from Part P and attach a 1/2" square nut, as shown in Figure 11-14. Use two 1 1/2" bolts and matching nuts (see Figure 11-15) to secure the square nut over the 1/2" hole. The heads of the bolts will clamp down on the surface of the square nut and hold it in place.

Figure 11-13. *Drill a hole through Part E using a 1/2" drill bit.*

Figure 11-14. *Use bolts to secure a 1/2" square nut to Part E.*

Figure 11-15. *Use nuts to tighten down and hold the square nut on Part E.*

Reattach Part E to Part P. Next, insert the x-axis lead screw you cut back in Chapter 9 through a lead screw bearing in one of the tabletop legs. Thread the lead screw through the square nut bolted to Part E, and then keep threading the lead screw until the opposite end exits the other tabletop leg's bearing.

TIP Threading the long lead screw through both legs and Part E can take a while; wearing gloves will make it a little easier on your hands.

Summary of Work

At this point, you should have the following items completed:

One BRA attached to Parts Q and R

Part P cut to proper length (distance between Parts Q and R when mounted to table)

Square nut bolted to Part E

Part E bolted to Part P

Part P bolted between Parts Q and R

Parts P, Q, and R assembled as a partial frame and mounted to the table

X-axis lead screw inserted through lead screw bearings and screwed through Part E

Hardware Required

For the work performed in this chapter, you will use

　　　1/4" bolts; 1 1/4" length; quantity: 12

　　　1/4" nuts; quantity: 16

　　　1/4" bolts; 3" length; quantity: 8

　　　1/4" bolts; 1 1/2" length; quantity: 2

　　　Cross dowels; quantity: 6

　　　1/2" square nut; quantity: 1

What's Next?

In Chapter 12, you're going to complete the y-axis frame by cutting and drilling two more pieces of MDF that will be attached above the table's surface and between Parts Q and R. When done, the y-axis frame should be sturdy and allow for smooth movement forward and backward on the table surface (x-axis).

Y-Axis, Part 3

Now that you have the gantry sides (Parts Q and R) bolted to the Gantry Bottom Support (Part P), it's time to finish up the y-axis frame by adding the Y-Axis Rail Support. Consisting of two pieces of MDF bolted together, these two pieces will provide stability and rigidity to your CNC machine.

The y-axis frame you're building performs two functions. It rolls along rails mounted on the sides of the table, allowing the machine's router to move forward and backward. It also provides a "minitable" that allows the z-axis frame (covered in Chapter 13) to move side to side.

The Y-Axis MDF Parts

The y-axis frame consists of six pieces of MDF:

Part O	Y-Axis Rail Support (front)
Part P	Y-Axis Gantry Bottom Support
Part Q	Y-Axis Gantry Side (completed in Chapter 10)
Part R	Y-Axis Gantry Side (completed in Chapter 10)
Part S	Y-Axis Rail Reinforcement (back)
Part E	Y-Axis Gantry Bottom Nut

In this chapter, you'll be cutting, drilling, and mounting Parts O and S—these two pieces are bolted together and mounted above the tabletop and between the two gantry sides. Figure 12-1 shows Part O outlined (Part S is bolted to the back of Part O and not visible in the figure).

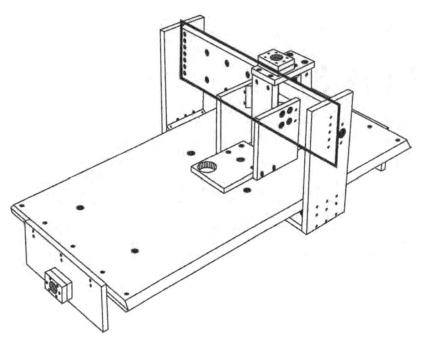

Figure 12-1. *Parts O and S are mounted between the gantry sides above the tabletop.*

NOTE Please refer to the MDF Parts Layout 1 and MDF Parts Layout 2 PDF files available for download at www.buildyourcnc.com/book.aspx for part names and letters. Refer to the MDF Plans and Cut List PDF file for cutting and drilling dimensions of all MDF parts—this file can also be downloaded at www.buildyourcnc.com/book.aspx.

Please read through this entire chapter first before you begin cutting and drilling. Remember, measure twice (or more) and cut once.

The Rail Support

If you've completed all the work on your CNC machine up to Chapter 11, you should have a partial assembly like the one shown in Figure 12-2.

Refer back to Chapter 11 for the distance you measured between the two gantry sides. You're going to use this same measurement as the length for cutting Part O. The CNC plans provide a length of 26 11/16" (26 and 11/16 inches) and a height of 8". Substitute your measurement for the inside distance between Parts Q and R that you wrote down in Chapter 11. Cut Part S to the dimensions specified in the CNC plans.

Part O needs to be chamfered to hold the angled aluminum rail that will be bolted to the top and bottom edges. Figure 12-3 shows one of Part O's edges marked for chamfering and the router chamfer bit set to the proper depth. (You can see Part O with its edges chamfered in Figure 12-4.)

Figure 12-2. *The y-axis frame is partially completed.*

Figure 12-3. *Chamfer 1/4" from the front and back sides of Part O.*

Next, measure and mark all the holes for drilling on Part O. For the 7/16" cross dowel holes on the left and right sides of Part O, we suggest drilling 1/8" pilot holes. Figure 12-4 shows Parts O and S clamped together (and you can see the pilot holes drilled in Part O). Notice that Part S is centered on Part O; try to be as exact as possible before clamping the two pieces together.

TIP It's also beneficial to label the surfaces of Parts O and S that are touching so that when you bolt the two pieces together later, the holes will line up perfectly. If you forget to do this and you accidentally flip Part S over, you can always put the two pieces together and look to see if the pilot holes match up with Part O.

Figure 12-4. *Parts O and S are clamped together for drilling.*

Part S will be bolted to Part O, so you can drill the ten pilot holes through both parts to help ensure the bolt holes will match up exactly. Figure 12-5 shows the ten pilot holes drilled into Part O and through Part S (on the back).

You'll be using 1" bolts to hold Parts O and S together. When a bolt and nut are tightened, both the bolt head and the nut must be below the surface of Parts O and S. To do this, you'll need to counterbore the surface of both pieces. Figure 12-6 shows a few of the counterbore holes drilled using a 1" Forstner bit. The approximate depth drilled into Part O is between 1/4" and 3/8"—you'll need to experiment with your counterbore bit to determine the proper depth for ensuring that the bolt head is below the surface of Part O.

Figure 12-5. *Ten pilot holes are drilled into Part O and through Part S (behind Part O).*

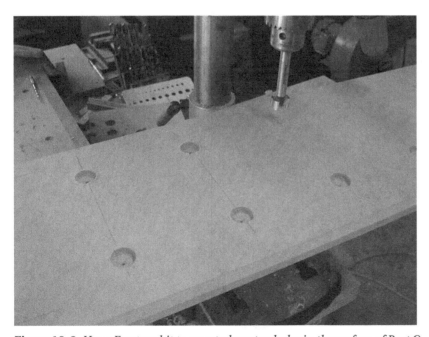

Figure 12-6. *Use a Forstner bit to counterbore ten holes in the surface of Part O.*

Perform the same actions on Part S; counterbore 10 holes an approximate depth of 1/4" to 3/8" so that the nut sits below the surface of Part S. Figure 12-7 shows a few of the holes counterbored for Part S.

Figure 12-7. *Counterbore the holes on Part S that will hold the nuts used to bolt Part S to Part O.*

Next, use a 1/4" drill bit to drill holes (for the 1" bolts) in all 20 of the counterbore holes (on Parts O and S). Figure 12-8 shows these holes being drilled into Part O.

Figure 12-8. *Drill 1/4" holes into Parts O and S for the 1" bolts.*

Next, use a 7/16" bit to drill the holes on the left and right sides of Part O that will hold the cross dowels. Also drill the 1/4" holes into Part O's edges where the bolts will be inserted and screwed into the cross dowels. Figure 12-9 shows the cross dowel holes being drilled.

Figure 12-9. *Drill the holes for the cross dowels that will be inserted into Part O.*

Figure 12-10 shows the 1/4" holes drilled into the edges of Part O.

Figure 12-10. *Drill holes in the edges for bolts to screw into the cross dowels.*

Next, cut two lengths of angled aluminum rail to match the length of Part O. Figure 12-11 shows the rail placed on top of Part O but not yet bolted down.

Figure 12-11. *Cut two pieces of rail to be bolted to the top and bottom of Part O.*

Now it's time to bolt all the pieces together and attach them to your CNC machine.

Finishing the Y-Axis Frame

Insert ten 1" bolts and washers into the counterbored holes on Part O, as shown in Figure 12-12. Push these bolts into Part O and through Part S.

Attach ten nuts to the ten bolts inserted into Part O and through Part S. Figure 12-13 shows a nut tightened down; note that it sits slightly below the surface in the counterbored hole.

Figure 12-12. *Bolt Parts O and S together using 1" bolts and washers.*

Figure 12-13. *Tighten down the nuts to hold Parts O and S together.*

Next, use ten 3" bolts (five per side) and ten cross dowels to attach Part O/S to the gantry sides, as shown in Figure 12-14.

NOTE Figure 12-14 shows Part S and the nuts used to secure it to Part O. It doesn't matter which way you attach Part O/S, but the direction Part O is facing will be the front of the CNC machine. Keep this in mind for future reference.

Figure 12-14. *Parts O and S bolted to the gantry sides*

Place the aluminum rails on the top and bottom of Part O/S, as shown in Figure 12-15. We used painter's tape to hold the two pieces in place.

Mark the aluminum for drilling; you'll be drilling two holes into the aluminum so that 2" bolts can be inserted and screwed into two cross dowels, one on each end of the rail. Drill the holes into the rail and bolt them to Part O/S. Figure 12-16 shows the rails bolted to the frame.

Figure 12-15. *Use tape to hold the rail in place so you can mark it for drilling.*

Figure 12-16. *The aluminum rail drilled and bolted to the Y-Axis Rail Support (Part O).*

And that's it for the y-axis frame! You're now approximately two-thirds done with your CNC machine. Take a break and have some fun sliding the frame forward and backward along the tabletop's rails.

Tips on Final Frame Assembly

After you've bolted on Part O/S, you may find that the y-axis frame grips the tabletop rails either too tightly or too loosely. Here are a few suggestions for ensuring that the BRAs roll smoothly on the rails.

- *BRAs are too tight on the rails, part 1*: This can be caused by the gantry sides and their matching BRAs pressing too tightly against the x-axis rails they will ride on. To alleviate this problem, you can insert washers in the gaps between the gantry sides and Part O and/or Part P to move the gantry sides further out (and by further, we're talking about maybe a 1/16" distance). This will, in turn, move the BRAs slightly away from the rail, reducing pressure and providing a smoother motion. You'll have to experiment with different washer placements.

- *BRAs are too tight on the rails, part 2*: This can also occur if Part O and Part P are cut too short in length. Once again, you can insert washers in the gaps between the gantry sides and Part O and/or Part P to move the gantry sides further out. But if Parts O and P are cut too short, no amount of washers will really help, and you may be forced to remeasure the distance between the gantry sides and recut Parts O and P.

- *BRAs are too loose on the rails*: If you cut the length of Parts O and P too long, the gantry sides will not hold the BRAs securely to the rail. Grab the y-axis frame and wiggle it a bit—do you hear rattling from the BRAs? One solution is to trim Parts O and P down just a bit—maybe just a 1/32" or 1/16" on each side. This will pull the gantry sides in closer, and the BRAs will more securely ride the rails. (Of course, if the BRAs then become too tight, refer to the preceding two points for using washers to fix this problem.)

Summary of Work

At this point, you should have the following items completed:

Parts O and S drilled

Parts O and S bolted together

Two pieces of angled aluminum rail cut, drilled, and mounted to Part O/S

Part O/S with rails mounted to the y-axis frame

Hardware Required

For the work performed in this chapter, you will use

1/4" bolts; 1" length; quantity: 10

1/4" nuts; quantity: 10

1/4" bolts; 3" length; quantity: 10

1/4" bolts; 2" length; quantity: 4

Cross dowels; quantity: 14

What's Next?

Chapter 13 is where you're going to begin building the z-axis assembly. This assembly will move side to side along the y-axis frame's rails and provide the machine's router with the ability to move up and down with respect to the tabletop surface. You'll be building some more BRAs, cutting rail, and drilling . . . so let's get started.

Preparing for the Z-Axis

You should now have a better understanding of how the various axes are going to be used to get the machine's router bit to the material for cutting and drilling. Up to this point, you've built the x-axis and y-axis, including the y-axis frame that will move forward and backward on the x-axis aluminum rails.

You should also be much more comfortable with cutting, drilling, and measuring MDF, as well as creating the very useful bearing-rail assemblies (BRAs). Well, you're almost done.

The z-axis frame is next. This frame will ride side to side along the y-axis frame you put together in Chapter 12. In addition, this frame will contain another set of rails, mounted vertically, that will allow the router to move up and down. Once this frame is built and attached, you'll have the makings of a real CNC router with full motion on all three axes.

We know you're anxious to complete your CNC machine, so we'll jump right into it. Before we begin building the z-axis, however, there are two more parts related to the y-axis that must be cut. These two parts are needed to make a critical key measurement for cutting and assembling the final z-axis frame.

The Y-Axis BRA Supports

In order to mount your z-axis frame to the y-axis frame, you'll be cutting and drilling two parts that make up the top and bottom "caps" of the z-axis frame.

The two MDF parts you will be cutting are

Part C   Y-Axis Linear Bearing Support—Top (BRA Support)

Part D   Y-Axis Linear Bearing Support—Bottom (BRA Support)

NOTE Please refer to the MDF Parts Layout 1 and MDF Parts Layout 2 PDF files available for download at www.buildyourcnc.com/book.aspx for part names and letters. Refer to the MDF Plans and Cut List PDF file for cutting and drilling dimensions of all MDF parts—this file can also be downloaded at www.buildyourcnc.com/book.aspx.

Figure 13-1 shows these two parts outlined; Part D is partially obscured in the figure because it is on the bottom of the z-axis frame.

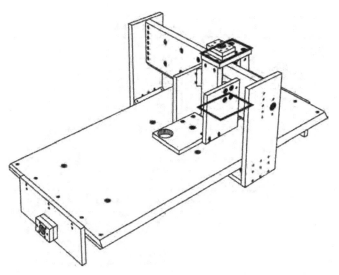

Figure 13-1. *Parts C and D are the caps to the z-axis frame and hold the BRAs.*

Please note that Part C has a single BRA mounted on the underside that rides along the top y-axis frame rail. Part D has a matching BRA that is mounted on the top that rides the bottom y-axis frame rail.

Cutting and Drilling Parts C and D

Figure 13-2 shows a portion of the MDF Parts Layout 1 PDF file. You can see that Parts C and D are side by side.

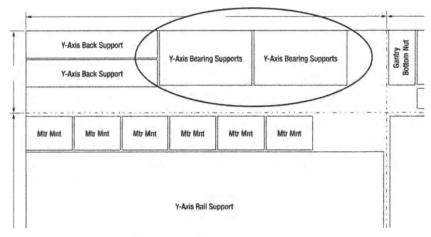

Figure 13-2. *Parts C and D cut out of MDF*

The dimensions of Parts C and D are fixed; cut these pieces to 4"×6 7/8" (6 and 7/8 inches).

You're also going to need two BRAs after cutting and drilling Parts C and D. So, go ahead and cut two pieces of angled aluminum rail to a length of 4" each. Use your BRA template to drill pilot holes and then tap both pieces to make two BRAs. Set these aside until you're done drilling Parts C and D.

Figure 13-3 shows Parts C and D and two 4" lengths of rail.

Figure 13-3. *Two 4" rails and Parts C and D cut and waiting to be marked for drilling*

Tape Parts C and D together (we used blue painter's tape) and mark the top piece for drilling using the dimensions provided in the CNC plans. Because Parts C and D will be drilled identically, drilling pilot holes through both pieces first will ensure that the holes in both parts match up later. For now, do not drill the larger hole or the two small 1/4" holes above and below it, as indicated in Figure 13-4. That will be covered in a later chapter. We used a 3/4" Forstner bit for the larger counterbore holes and a 1/2" Forstner bit for the smaller holes.

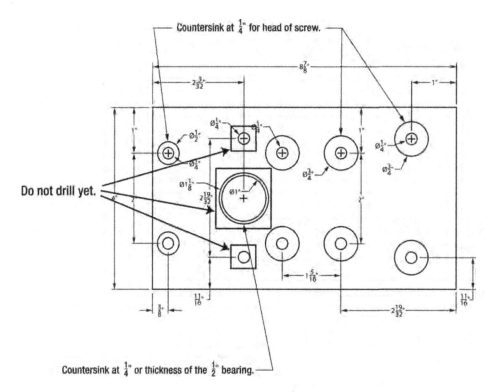

Figure 13-4. *Counterbore the holes as indicated in the plans, but do not drill the larger hole.*

Figure 13-5 shows Parts C and D taped together and marked for drilling. The goal is to drill using a 1/8" bit to create pilot holes. Note that the blue painter's tape is applied over the spot where the larger hole will be drilled. Go ahead and drill the pilot holes in Parts C and D.

CAUTION After taping, label both of the exposed surfaces with "CB" to indicate the surface where the counterbored holes will be drilled. (The surfaces of Parts C and D that are touching and hidden between the two taped parts will not have counterbored holes drilled, but will instead hold the BRAs.)

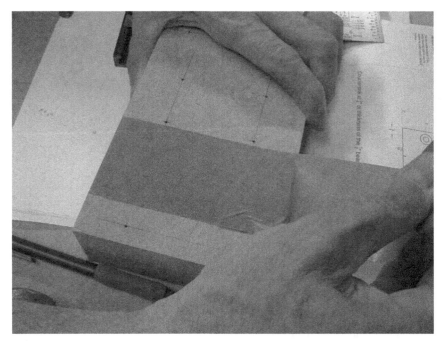

Figure 13-5. *Tape Parts C and D together for drilling.*

After drilling the pilot holes, remove the tape to drill the counterbore holes.

TIP The surfaces of Parts C and D with the counterbored holes will face away from the z-axis frame. This means that Part C (the top) will have the counterbore holes facing up, away from the machine. Part D's counterbored holes will be facing down. This is why you need to label each surface with "CB" so you don't forget which side is which.

Figure 13-6 shows us drilling the six counterbored holes in Part C. You'll need to experiment with the proper depth so that the bolt heads that go into the counterbored holes sit completely below the surface. For our Part C, the counterbore depth was approximately 1/4".

Drill both Parts C and D with the counterbored holes. Figure 13-7 shows the two pieces of MDF with all the counterbore holes drilled.

Figure 13-6. *Drilling the counterbored holes in Part C*

Figure 13-7. *Parts C and D with counterbore holes drilled*

Notice in Figure 13-7 that the center of the counterbored holes still have the 1/8" pilot holes. Use these as guides to drill 1/4" holes in the center of all the counterbored holes. Figure 13-8 shows Parts C and D with the 1/4" holes drilled.

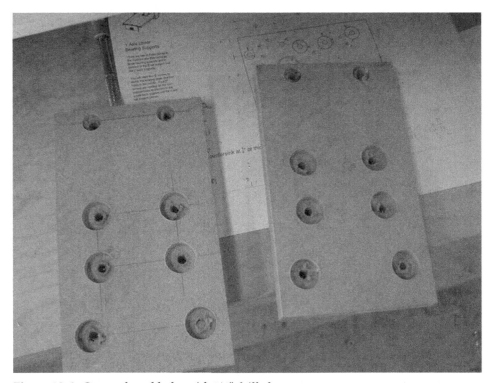

Figure 13-8. *Counterbored holes with 1/4" drilled centers*

Next, mount one BRA each on Parts C and D. Mount the BRAs on the side that does not have the counterbored holes drilled. Figure 13-9 shows the BRAs mounted.

Figure 13-9. *BRAs mounted to Parts C and D*

You've now completed the two caps that will hold the z-axis frame to the y-axis frame. All that's left is to drill the large hole in both parts to hold the lead screw bearing. Refer to the plans for the locations for drilling the large bearing holes (and for Part C, the two 1/4" holes that will hold the square nut). Drilling these large holes is identical to the method described in Chapter 10. You'll use a 1 1/8" Forstner bit to drill the larger hole and a 1" bit for the inner hole. Figure 13-10 shows Part D with the lead screw bearing hole drilled. Note that the counterbored hole is on the same side as the counterbored holes for the nuts that hold the BRAs.

Figure 13-10. *Drilling the lead screw bearing hole in Part D*

Figure 13-11 shows Part C with the bearing inserted along with a test piece of lead screw. After drilling the lead screw bearing holes for Parts C and D, set these parts aside.

Figure 13-11. *A piece of lead screw will go through the bearing inserted in Part C.*

Before we finish building the z-axis, we've got to take a measurement using Parts C and D. This measurement is critical when it comes to the assembly and movement of the z-axis, so take your time and maybe even have someone else double-check your numbers.

Measuring for the Z-Axis

There are three parts that you're going to cut in the next few chapters; these are Parts A, B, and F. Two of these parts (F and A) are indicated in Figure 13-12; Parts F and A are partially visible, but Part B is completely obscured in the figure. Part B is mounted vertically next to Part A and has the same dimensions as Part A.

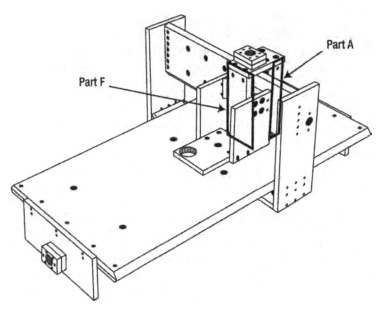

Figure 13-12. *The height of Parts A, B, and F must be measured accurately.*

Figures 13-13 and 13-14 show the final parts installed. Ignore the assembly for now and just focus on the distances that must be measured, as indicated in each figure.

Figure 13-13. *Parts A and B on the back of the CNC machine*

Figure 13-14. *Part F will face the front of the CNC machine.*

Now here's the tricky part—we highly recommend an assistant for this task. Place Part C on top of the y-axis frame's aluminum rail and place Part D under the rail. This task is much easier if you use two clamps—one in the front and one in the back—as shown in Figure 13-15.

Figure 13-15. *Clamp Parts C and D to the rail and measure the distance.*

It will help if you use a small (torpedo) level to make certain that Parts A and B are as parallel to the floor (or tabletop surface) and each other as possible. Adjust the clamps until the distance on the front (for Part F) matches the distance you've measured on the back (for Parts A and B). Once you have Parts C and D clamped to the rails and the front and back measurements are equal, write down that measurement here:

Height/length of Parts A, B, and F: _____

Summary of Work

At this point, you should have the following items completed:

Part C and D cut and drilled

Two BRAs assembled using a 4" rail length

One BRA bolted to Part C and one bolted to Part D

Parts C and D clamped to the y-axis frame rail for measuring

Height/length of Parts A, B, and F measured

Hardware Required

For the work performed in this chapter, you will use

    Bearings; 1/2" inner diameter, 1 1/8" outer diameter; quantity: 2 (for lead screw)

    Bearings; 5/16" inner diameter, 7/8" outer diameter, 1/4" thick; quantity: 8 (for BRAs)

    1/4" bolts; 1" length; quantity: 8 (for BRAs)

    1/4" nuts; quantity: 16 (for BRAs)

What's Next?

Up next in Chapter 14, you'll begin cutting the parts required to build the z-axis frame. You'll be building another set of BRAs and cutting Part F and chamfering its edges to hold two pieces of aluminum rail.

Z-Axis, Part 1

You're almost done with the cutting and drilling of MDF. The remaining part of the CNC machine that you're going to need to assemble is the z-axis frame. The z-axis frame will not only hold the router you'll install, but it will also give it side-to-side and up-and-down mobility. (The front-to-back motion is supplied by the y-axis frame that rides on the x-axis tabletop.)

As with earlier tasks, you'll be doing some more cutting and assembling using cross dowels, as well as building some more BRAs. You'll also be using some measurements you took back in Chapter 13, so be sure you've followed the instructions in that chapter carefully.

The Z-Axis MDF Parts

For your z-axis, you're going to need to cut a total of six pieces. The MDF parts you will be cutting are

Part F	Z-Axis Rail Support
Part M	Z-Axis Back Support
Part N	Z-Axis Back Support
Part W	Z-Axis Bearing Support
Part X	Z-Axis Bearing Support
Part V	Router Base

NOTE Please refer to the MDF Parts Layout 1 and MDF Parts Layout 2 PDF files available for download at www.buildyourcnc.com/book.aspx for part names and letters. Refer to the MDF Plans and Cut List PDF file for cutting and drilling dimensions of all MDF parts—this file can also be downloaded at www.buildyourcnc.com/book.aspx.

Again, we highly encourage you to read through this entire chapter first before you begin cutting. Why? So you'll understand why we're recommending that you wait to cut Parts M, N, and V until you actually need them in the process.

So, the first three parts you'll be cutting and working with are Parts F, W, and X. We'll start with Part F.

Part F: The Z-Axis Rail Support

Figure 14-1 shows a portion of the MDF Parts Layout 1 PDF file. You can see that Parts E and F are side by side.

CAUTION The part layouts found in MDF Parts Layout 1 and MDF Parts Layout 2 *are not to scale*, refer to the MDF Plans and Cut List file for actual dimensions.

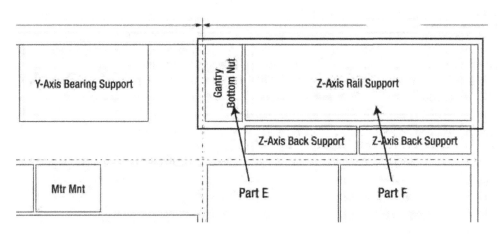

Figure 14-1. *Part F is on the right side and its dimensions are 4"×10 1/16".*

Figure 14-2 shows Part E and Part F after being cut. (Part E was used in Chapter 11 and bolted to the bottom of the CNC machine.) Part F has unique dimensions. Back in Chapter 13, you measured the distance between Parts C and D. Refer back to that chapter and transfer the measurements you took for Parts A, B, and F here:

Height/length of Parts A, B, and F: _____

Cut Part F to its correct dimensions of 4" wide and a height equal to the measurement recorded previously.

Figure 14-2. *Part E (used earlier in Chapter 11) and Part F cut out of MDF*

The long sides of Part F need to be chamfered. Figure 14-3 shows a standard-sized router with the chamfer bit installed. Make note of the two small pencil lines on the edge of Part F (also shown in Figure 14-3).

Figure 14-3. *A chamfer bit is required for the edges of Part F.*

Look at Figure 14-4, which shows what Part F will look like after both edges are chamfered. You're basically cutting a 45 degree corner along each of the long sides of Part F.

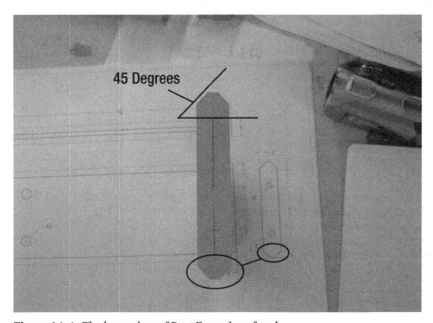

Figure 14-4. *The long edges of Part F are chamfered.*

You'll run your router with a chamfer bit down the long edges of Part F on top and bottom. You want to leave 1/4" of flat surface, so make pencil marks (like the ones shown in Figure 14-3) and set your chamfer bit to cut away enough MDF material to leave a 1/4" surface, as shown in Figure 14-5. Because the MDF is 3/4" thick, you'll simply make two marks and divide the edge into three parts of 1/4" each.

Chamfer both long edges on the top and bottom of Part F. Figure 14-6 shows Part F with the chamfering completed.

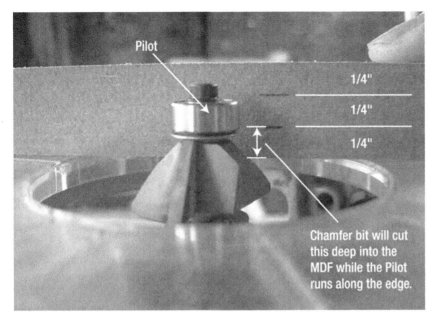

Figure 14-5. *Set the chamfer bit to cut away only 1/4" of an edge.*

Figure 14-6. *Part F with top and bottom long edges both chamfered*

Set Part F aside for now. You'll be marking and drilling some holes in it a bit later.

Parts W and X: The Z-Axis Bearing Supports

The next two parts you'll be cutting out are Parts W and X, or the Z-Axis Bearing Supports. There are two of them, each 7"×8 13/32" in size. These parts are found in the MDF Parts Layout 2 PDF file; a portion of that file is shown in Figure 14-7.

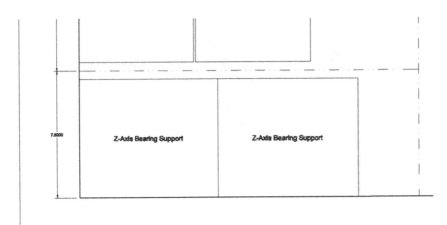

Figure 14-7. *The two Z-Axis Bearing Supports to be cut from the MDF*

Figure 14-8 shows where these two pieces are located in the final CNC machine.

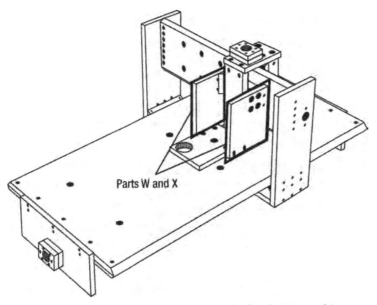

Figure 14-8. *Location of Parts W and X on the final CNC machine*

Finally, Figure 14-9 shows Parts W and X cut out from MDF. Remember to label each part that you cut—something as simple as writing the part letter in a corner will suffice, as shown in the figure.

Figure 14-9. *Parts W and X cut and trimmed to proper size*

You'll be measuring and drilling holes in these parts in Chapter 15, but for now go ahead and set these two parts aside. You've got one more task to complete before finishing this chapter.

Z-Axis Bearing-Rail Assemblies

You're going to need to build two z-axis bearing-rail assemblies (BRAs). The instructions for doing this are found in Chapter 4. Cut two small pieces of 3/4" aluminum rail with a length of 4". Figure 14-10 shows these two pieces cut.

Next, you'll drill eight pilot holes (four in each piece) in the rails. Use these pilot holes to then drill the 17/64" holes. (Refer to Chapter 4 for complete instructions.) Add threads to the holes using the tap tool and then screw four of the bolt-bearing-nut assemblies into each rail. Figure 14-11 shows the two z-axis BRAs fully assembled and ready.

Figure 14-10. *Cut two pieces of aluminum rail for the z-axis BRAs.*

Figure 14-11. *Two BRAs for the z-axis*

Summary of Work

At this point, you should have the following items completed:

Part F cut to 4" wide and a custom length that you measured in Chapter 13

Part W with dimensions of 7"×8 13/32"

Part X with dimensions of 7"×8 13/32"

Part F chamfered on both sides and along both long edges (10 1/16" edges)

Two BRAs, each 4" in length

Hardware Required

For the work performed in this chapter, you will use

Bearings; 5/16" inner diameter, 7/8" outer diameter, 1/4" thick; quantity: 8 (for BRAs)

1/4" bolts; 1" length; quantity: 8 (for BRAs)

1/4" nuts; quantity: 16 (for BRAs)

What's Next?

Up next in Chapter 15, you're going to do some measuring, marking, and drilling using Parts F, W, and X. The most important part of the work will be accurately measuring the spots where the drilling will occur.

Z-Axis, Part 2

In Chapter 14, you cut a few of the parts required to build the z-axis. In this chapter, you're going to drill some holes in a few of those pieces, take a few measurements, and cut the remaining parts required for the z-axis.

As you read this chapter, keep in mind that a few of the parts you'll use for assembling the z-axis may not match exactly with the measurements we provide. This is not necessarily a mistake on your part; slight variations in cutting the MDF can cause some measurements to fluctuate as much as 1/16". Just be patient and keep moving forward, and your z-axis will begin to take shape.

Preparing to Drill

Before you can assemble your z-axis, you're going to need to drill some holes so the various pieces you have cut can be bolted together. Just go slow, consult the CNC machine plans so you can properly mark the spots on your MDF pieces where you'll be drilling, and then select the properly sized drill bit for your drill.

The MDF parts you will be drilling in this chapter are

Part F   Z-Axis Rail Support

Part W   Z-Axis Bearing Support

Part X   Z-Axis Bearing Support

NOTE Refer to the MDF Plans and Cut List PDF file for cutting and drilling dimensions of all MDF parts—this file can be downloaded at www.buildyourcnc.com/book.aspx.

We'll begin drilling with Part F.

Drilling Part F: The Z-Axis Rail Support

Figure 15-1 shows the undrilled Part F. Back in Chapter 14 we suggested that you mark the piece for drilling before chamfering the edges. If you have not marked Part F for drilling, go ahead and consult your plans and mark the four points where you'll be drilling. You will be using a 7/16" drill bit to drill four holes that will hold four cross dowels.

Figure 15-2 shows Part F after being drilled. The center of each hole is 3/4" from the shorter end of Part F and 1" from the chamfered edge's widest point.

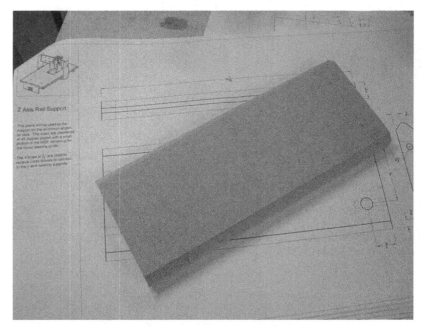

Figure 15-1. *Part F is chamfered and ready for drilling.*

Figure 15-2. *Part F with four drilled holes for cross dowels*

Notice in Figure 15-2 that the plans also call for 1/4" holes to be drilled into the sides of Part F, too. These holes should extend about 1/4" past the 7/16" holes that were drilled for the cross dowels; this will allow the 1/4" bolts to be screwed into and through the cross dowels.

Figure 15-3 shows how we're going to drill into Part F horizontally; all that's left is to clamp the piece down so it doesn't move. Go slow and make certain to keep the drilling as level and straight as possible. We used a small "bubble level" built into our drill to help with drilling accuracy.

Figure 15-3. *Drilling the 1/4" holes into the sides of Part F*

Drill all four holes and then set Part F aside for now. Our next step is going to be drilling Parts W and X.

Drilling Parts W and X: The Z-Axis Bearing Supports

Take a look at Figure 15-4 and you'll see a closeup of Part W with all the drill marks in pencil.

Now take a look at Figure 15-5. This is a portion of the plans for Part W and shows that some of the holes should be drilled with a Forstner (counterbore) bit. This will allow the bolt heads to be below the surface of the MDF.

Figure 15-4. *Part W with drill marks made in pencil*

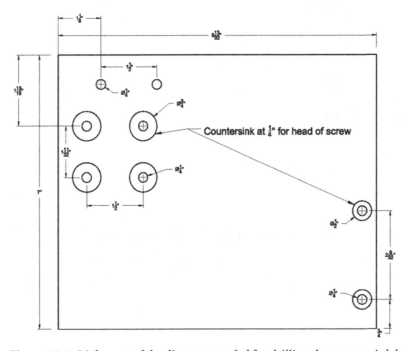

Figure 15-5. *Make note of the diameter needed for drilling the countersink holes.*

All of the smaller holes shown in Figure 15-5 are 1/4" holes. But before you drill those, we suggest that you drill the larger counterbore holes first. Figure 15-6 shows Part W partially completed—the larger counterbore holes have been drilled, and all that's left is for us to drill the 1/4" holes. Notice that the drill bit we used left a small dimple in the center of each hole; we'll use this dimple to help us center the 1/4" drill bit. (One of the single 1/4" holes has already been drilled in the upper-left corner.)

Figure 15-6. *Part W with the counterbore holes drilled*

NOTE We recommend drilling a test counterbore hole in a piece of scrap MDF to help you determine the proper depth to drill. For our Part W, we found that a depth of 1/4" was sufficient to allow the bolt head to sit just below the surface. Don't drill too deeply with the counterbore bit.

If you consult the plans carefully, you'll notice that Part X is an exact mirror of Part W. Take a look at Figure 15-7 and you can see that Parts W and X are drilled almost identically, but the counterbore holes are on the outward-facing surfaces (refer back to Figure 14-8 and you can see the counterbore holes on Parts W and X face away from the center of the z-axis.)

Figure 15-7. *Parts W and X drilled with counterbore holes drilled on opposing sides*

That's all the drilling for now. Up next, we need to mount the bearing-rail assemblies (BRAs) we made in Chapter 7 to the sides of Parts W and X.

Mounting the Z-Axis Bearing-Rail Assemblies

Grab the two BRAs you made back in Chapter 7 and flip over Part X to the side with no counterbore holes. Figure 15-8 shows how we've bolted down a BRA using standard 1/4" bolts.

You might find it easier to bolt down the BRA by removing the four bolt-bearing-nut assemblies. Tighten the 1/4" bolts down a little at a time and don't overtighten. When done, put the bolt-bearing-nut assemblies back on. Do this for Parts W and X. Figure 15-9 shows the BRAs added to Parts W and X.

Figure 15-8. *One of the BRAs bolted down to Part X*

Figure 15-9. *Location of Parts W and X on the final CNC machine*

And that's it for Chapter 15. You're almost done with the z-axis!

Summary of Work

At this point, you should have the following items completed:

Part F with eight holes drilled; four on surface for cross dowels and two on each end

Part W drilled (along with countersunk holes) with BRA mounted

Part X drilled (along with countersunk holes) with BRA mounted

Hardware Required

For the work performed in this chapter, you will use

1/4" bolts; 1" length; quantity: 8 (for attaching BRAs)

1/4" nuts; quantity: 8

What's Next?

You're going to complete the z-axis assembly in Chapter 16. This will involve assembling Parts F, W, and X. But don't worry—it's not hard to do. And when you're done, you'll have 90 percent of your CNC machine completed!

Z-Axis, Part 3

In Chapters 14 and 15 you cut and drilled three of the parts required to build the z-axis and then mounted two bearing-rail assemblies (BRAs). In this chapter, you're going to finish the assembly of the z-axis using the custom measurements you took back in Chapter 13 and another custom measurement you'll perform shortly.

These custom measurements are going to be used to cut the remaining pieces for the z-axis. Please read through the entire chapter before beginning so you'll have a better understanding of the work that you'll be doing.

NOTE Refer to the MDF Plans and Cut List PDF file for cutting and drilling dimensions of all MDF parts—this file can be downloaded at www.buildyourcnc.com/book.aspx.

Cutting the Z-Axis Rail Support Rails

Before you perform a test assembly of the current z-axis parts, you need to cut two pieces of 3/4" angled aluminum rail for the chamfered edges of Part F. The lengths of these two pieces will be 10 1/16" (10 and 1/16 inches). Figure 16-1 shows the two pieces of rail cut for our Part F.

Figure 16-1. *Part F with two matching rails cut from 3/4" angled aluminum*

You won't need to drill into these rails; you'll see shortly how they will be held tightly to the sides of Part F by the two BRA you mounted to Parts W and X in Chapter 15.

Take a good look at Figure 16-2. You can see that Parts W and X are standing vertically and their attached BRAs are holding the rail to the sides of Part F. You'll need to accurately measure the distance between Parts W and X. To do this, we recommend using a set of clamps to hold the entire assembly together, as shown in Figure 16-2, but a helper can also be used to hold the collection of pieces together as you measure. Just make certain that Parts W and X are perfectly parallel to one another and that Part F can move up and down without any resistance. The entire assembly should be tight . . . but not so tight that the BRAs do not allow for easy up-and-down movement of Part F.

Figure 16-2. *Measure the distance between Parts W and X.*

The distance you measure between Parts W and X will be used to cut Parts M and N. Take a look at Figure 16-3. It shows a fragment of the CNC plans for Parts M and N (Z-Axis Back Supports) and their respective measurements. Did you notice that Part M (the top piece with five holes on its surface) and Part N (the bottom piece with three holes) do not have lengths attached to them? They are 1 1/2" wide, but the length will be equal to the distance you measured between Parts W and X.

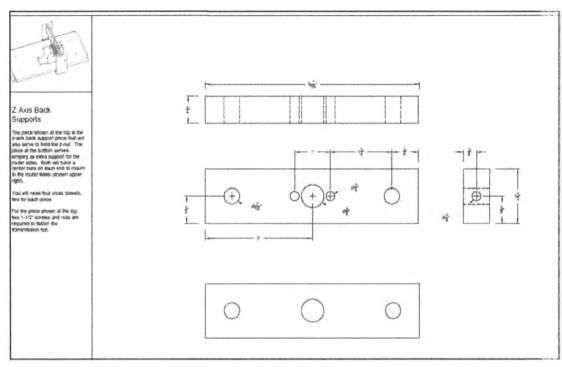

Figure 16-3. *Parts M and N will have a custom length that you will need to measure.*

Go ahead and cut Parts M and N using the dimensions provided in the plans.

Cutting and Drilling Parts M and N

Figure 16-4 shows Parts M and N cut to their proper lengths and widths, and we've marked the holes for drilling on each as shown in the plans.

You're going to be drilling a few different-sized holes in these two pieces, so we started by drilling the 7/16" holes that will be used to hold the cross dowels. You can see in Figure 16-5 that we've drilled two holes in Parts M and N.

Figure 16-4. *Part N (top) and Part M (bottom) cut and marked for drilling*

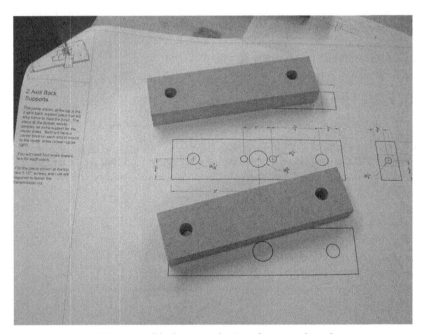

Figure 16-5. *Drill two 7/16" holes in each piece for cross dowels.*

Next, we drilled one 1/2" hole in each piece, as shown in Figure 16-6. These holes need to be drilled as precisely as possible, so check your measurements again and try to get the drill bit centered directly on your mark.

TIP After determining the center point where we will drill, we would use a hammer and a small nail to make a "dimple" on the spot to be drilled. This can help you when you're trying to get the drill bit on the exact center point, as the drill bit tip will tend to find the dimple and let you drill precise holes.

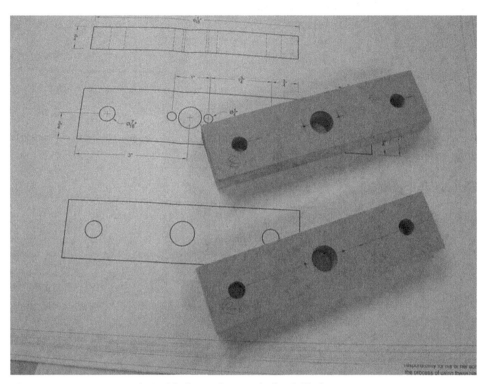

Figure 16-6. *Parts M and N with the 1/2" center holes drilled*

Finally, Part M requires two additional holes to be drilled at 1/4" diameter. Make certain when you're drilling these holes to keep the drill going into the MDF as straightly as possible so it doesn't angle into the larger 1/2" center hole. Figure 16-7 shows Parts M and N with the drilling completed.

All that's left at this point is to drill a 1/4" hole into the sides of Parts M and N for the bolts that will screw into the cross dowels. Go ahead and drill those holes; remember to go a little past the 7/16" hole that will hold the cross dowel so the bolt has space to screw into and through the cross dowel. Figure 16-8 shows two of the holes drilled into Parts M and N. Matching holes are also drilled on the opposite sides.

Figure 16-7. *Parts M and N with all the surface holes drilled*

Figure 16-8. *Parts M and N with the 1/4" side holes drilled for bolts*

Cutting and Drilling Part V

Figure 16-9 shows a partial view of the CNC plans for Part V, the Router Base. This is the base plate where you will be mounting the router for your CNC machine. The length of the part is 8", but the width will be equal to the measurement you made earlier for the lengths of Parts M and N. The Router Base will be bolted between Parts W and X, so its width may vary slightly due to the slight gap between the BRAs and the rails they ride on.

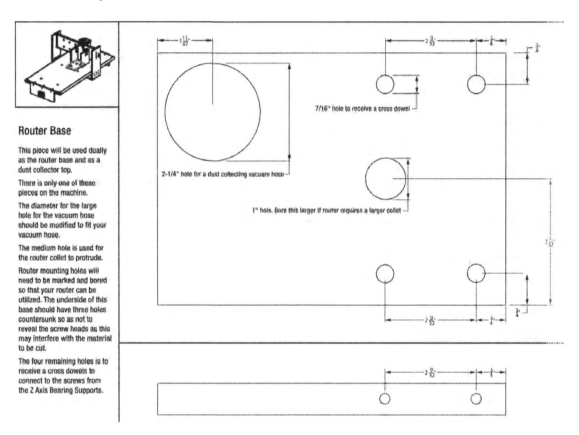

Router Base

This piece will be used dually as the router base and as a dust collector top.

There is only one of these pieces on the machine.

The diameter for the large hole for the vacuum hose should be modified to fit your vacuum hose.

The medium hole is used for the router collet to protrude.

Router mounting holes will need to be marked and bored so that your router can be utilized. The underside of this base should have three holes countersunk so as not to reveal the screw heads as this may interfere with the material to be cut.

The four remaining holes is to receive a cross dowels to connect to the screws from the Z Axis Bearing Supports.

Figure 16-9. *Part V, the Router Base, is the sixth piece needed for assembling the z-axis.*

Go ahead and cut this piece and label it Part V. For now, however, the only holes you need to drill on it are the four 7/16" holes that will hold cross dowels and the matching 1/4" holes drilled into the sides of Part V for the bolts that will screw into the cross dowels.

(The large hole in the upper-left corner of Figure 16-9 is for inserting a vacuum nozzle to vacuum up the dust created by your CNC machine. The smaller hole near the center of Part V is for your router, and this hole size may vary. We'll cover this later in Chapter 17.)

After you've cut and drilled Part V, you should have six pieces of MDF, two pieces of rail (for Part F), and two BRAs already mounted to Parts W and X. These parts are shown in Figure 16-10.

Figure 16-10. *The parts for the z-axis are ready for assembly.*

Assembling the Z-Axis

Assembling the z-axis by yourself can be a little tricky; if you've got someone who can help you here, call them over. You're going to start by clamping Parts W and X so that they hold the two pieces of angled rail to the sides of Part F, as shown earlier in Figure 16-2. You can insert Part V between Parts W and X to provide more stability, but don't bolt Part V in yet. Insert Part M horizontally, as shown in Figure 16-11, and use two 2" bolts and two cross dowels to secure it. (Note in Figure 16-11 that you can just see a little piece of Part V behind Part F and wedged between Parts W and X.)

Figure 16-11. *Use bolts and cross dowels to secure Part M horizontally between Parts W and X.*

Perform the same steps on Part N and secure it below Part M, as shown in Figure 16-12.

Figure 16-12. *Use bolts and cross dowels to secure Part N horizontally between Parts W and X.*

Flip the z-axis assembly over and use cross dowels and bolts to attach Part V, as shown in Figure 16-13.

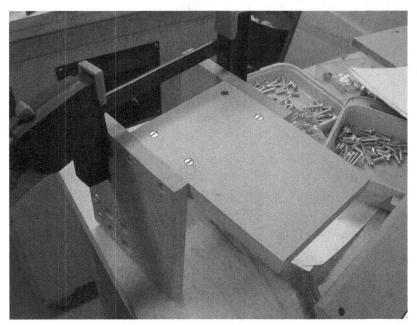

Figure 16-13. *Use bolts and cross dowels to secure Part V to the z-axis assembly.*

You can choose to leave Part V (the Router Base) unattached and set it aside for now. Since you'll be drilling it later for the vacuum and router holes described earlier, there's no need to bolt it to the z-axis. We attached it so we could take a picture of the final z-axis assembly shown in Figure 16-13.

Attaching the Z-Axis to the Machine

All that's left to connect your z-axis frame to the machine is cutting two more MDF parts: A and B, the Y-Axis Back Supports. Back in Chapter 13, you took a special measurement for the height/length of Parts A, B, and F. You also wrote that measurement down in Chapter 13. Refer back to Chapter 13, find that measurement, and write it down again here, because you're going to need to cut Parts A and B using that length now:

Height/length of Parts A, B, and F: _____

Figure 16-14 shows a portion of the CNC plans; Part A has a larger 5/8" hole near the surface's center and two smaller 1/4" holes surrounding it. Do not drill these three holes yet.

Part B also has the larger 5/8" hole in its center, but it does not have the small 1/4" holes. After cutting the two pieces, drill the 7/16" holes for cross dowels and 1/4" holes on the edges of Parts A and B for bolts to screw into cross dowels.

After cutting and drilling Parts A and B, your two pieces should look like the ones shown in Figure 16-15. Remember, the 5/8" holes and the smaller 1/4" holes will not be drilled until Chapter 17.

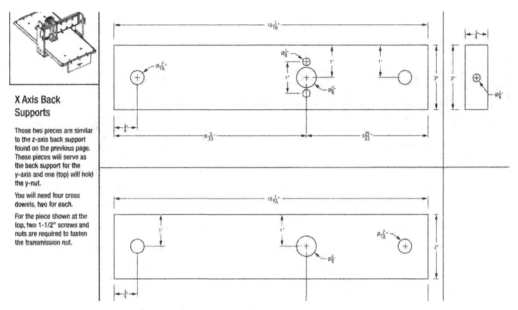

X Axis Back Supports

These two pieces are similar to the z-axis back support found on the previous page. These pieces will serve as the back support for the y-axis and one (top) will hold the y-nut.

You will need four cross dowels, two for each.

For the piece shown at the top, two 1-1/2" screws and nuts are required to fasten the transmission nut.

Figure 16-14. *Parts A and B are the Y-Axis Back Supports.*

Figure 16-15. *Parts A and B are almost identical, but not quite.*

Now you're ready to put these z-axis parts together. Take your z-axis frame and attach Parts C and D to Part F, as shown in Figure 16-16. Parts C and D, with BRAs attached, can be easily fitted over the rails attached to Part Y.

Figure 16-16. *Attach the z-axis frame to Parts C and D.*

Next, attach Parts A and B to Parts C and D. Figure 16-17 shows Parts A and B bolted into place.

Figure 16-17. *Attach Parts A and B to Parts C and D to complete the z-axis mounting.*

The z-axis frame should move easily side to side along Part Y. If it does not, this means that Parts C and D and the BRAs they are holding are pressing too tightly against the rail mounted to Part Y. An easy fix for this is to use washers (as shims) to "lengthen" Parts A and B. If you look carefully at Figure 16-18, you can see that a washer has been inserted between Part B and Part D. (A matching washer should also be added under Part A.) If you find that the BRAs are still too tight and not rolling smoothly, you can also place a washer on top of Parts A and B.

Once you've successfully bolted the z-axis frame to the CNC machine, you should have something that looks like Figure 16-19.

Figure 16-18. *Use washers to loosen BRAs that are too tight and not riding a rail smoothly.*

Figure 16-19. *Your CNC machine without any electronics*

We're not done yet with the machine, but you should be able to move all the parts smoothly on their rails.

Summary of Work

At this point, you should have the following items completed:

Rails cut to length for Part F

Parts M and N cut, drilled, and bolted to the z-axis frame

Part V cut, drilled, and bolted to the z-axis frame

Parts A and B cut and drilled

The z-axis frame bolted to the CNC machine using Parts A, B, C, D, and F

Hardware Required

For the work performed in this chapter, you will use

Cross dowels; quantity: 2 (Part M)

2" bolts; quantity: 2 (Part M)

Cross dowels; quantity: 2 (Part N)

2" bolts; quantity: 2 (Part N)

Cross dowels; quantity: 4 (Part V)

2" bolts; quantity: 4 (Part V)

Cross dowels; quantity: 2 (Part A)

2" bolts; quantity: 2 (Part A)

Cross dowels; quantity: 2 (Part B)

2" bolts; quantity: 2 (Part B)

What's Next?

Congratulations! You've completed the MDF frame of your CNC machine. . . some might argue the *most* important part since it's where the router and electronics (motors, drivers, etc.) will be added later. But we're not done yet. In Chapter 17, we're going finish the CNC frame by drilling and adding in the lead screws (refer to Chapter 6 for more information on lead screws) and customizing Part V to hold your router. After that, all that's left are the electronics and software.

Mounting the Electronics

You're almost done with the building and construction phase of your CNC machine. At this point, your machine has lead screws inserted through all three axes. You can turn these by hand (or use a drill) and watch as the individual axes move up and down, side to side, and front to back. But now it's time to bring your machine to life. Back in Chapter 6, you wired up the electronics and motors; in this chapter, you'll be mounting those motors to your machine.

This will be the last time you hear us say this, but please read through the entire chapter before beginning so you'll have a better understanding of the work that's ahead.

NOTE Refer to the MDF Plans and Cut List PDF files for cutting and drilling dimensions of all MDF parts—these files can be downloaded at www.buildyourcnc.com/book.aspx.

Cutting and Drilling the Motor Mounts

Your CNC machine will have three motors—one per axis. The motor will be bolted to the machine using a set of motor mounts, two per motor. Refer to the CNC plans for cutting and drilling these parts. Figure 17-1 shows a single motor mount cut and drilled with only the 2 1/2" large hole and the 1/4" holes used to bolt it to the CNC frame. (You'll be drilling four additional holes that will be used to mount the motor to the motor mount shortly.)

NOTE The motor mounts are Parts G, H, I, J, K, and L. Although all the motor mounts are cut and drilled identically, you will hear us refer to a motor mount as the "top motor mount" or the "bottom motor mount" because you will be using two motors mounts, stacked on top of one another, throughout the chapter. The top motor mount will be the motor mount that is closest to its attached motor; the bottom motor mount will, obviously, be the one underneath.

Figure 17-1. *A motor mount with the large hole drilled and two 1/4" holes*

Take three of the motor mounts (write *T* on them if you like to indicate "top") and place a motor on top of each of them so that the motor's shaft is centered in the large hole. Insert a drill bit (1/4") into the four corner holes on each motor and make four marks in the motor mount for drilling. Figure 17-2 shows each of the four corners of a motor being marked on a motor mount. You will be using #8 screws to bolt each motor to the motor mounts.

Clamp a bottom motor mount piece to a top motor mount and drill 1/4" holes through both pieces. Figure 17-3 shows a top motor mount and a bottom motor mount with the holes drilled to match the motor's four corners.

Figure 17-2. *Use the four corner holes of a motor to mark where to drill.*

Figure 17-3. *A top and bottom motor mount with corner holes drilled for motors*

Next, take the three bottom motor mounts and counterbore the four holes for mounting the motors so that the four #8 nuts you will use to secure each motor are flush with the surface (or slightly below). Figure 17-4 shows two of the bottom motor mounts—one with the #8 nuts inserted and the other with the holes counterbored. (We used a hammer and lightly tapped them into the holes. You want to make sure that they fit nice and tight so they won't spin.)

Figure 17-4. *Counterbore four holes in the bottom motor mount for #8 nuts.*

Next, place a top motor mount over a bottom motor mount and use 1/4" bolts (3" length) to secure the motor mounts to the CNC frame. Attach one motor mount pair to the front leg of the CNC machine (see Figure 17-5), another pair to one of the gantry sides (see Figure 17-6), and the remaining pair to Part C, the top of the z-axis (see Figure 17-7). After bolting on the motor mount pairs, attach a motor/lead screw coupling. Use an Allen wrench to tighten down one end of the couple to the lead screw—that end will have a 1/2" hole (the other end will eventually be tightened down on the motor shaft). Figures 17-5, 17-6, and 17-7 all show the motor mount pairs bolted and the motor/lead screw couples.

Figure 17-5. *A motor mount pair bolted to one of the x-axis supports (legs)*

Figure 17-6. *A motor mount pair bolted to one of the gantry sides*

Figure 17-7. *A motor mount pair bolted to the top of the z-axis*

Next, insert four #8 bolts (2 1/2" length) into the four corners of a motor. Screw two #8 nuts on to each bolt (don't tighten them, but screw them on until they're almost to the head of the bolt).

Take the motor, insert its shaft into the couple, and use an Allen wrench to tighten down the small screw to the shaft, as shown in Figure 17-8. (If you look carefully, you can see two nuts on each of the four #8 bolts.)

For each of the #8 bolts, use your fingers to screw down the nut closest to the motor mount, but don't tighten completely yet. Use your fingers to screw the nut closest to the motor up to the motor's bottom. Figure 17-9 shows this done for the z-axis motor; perform these same steps for the remaining two motors.

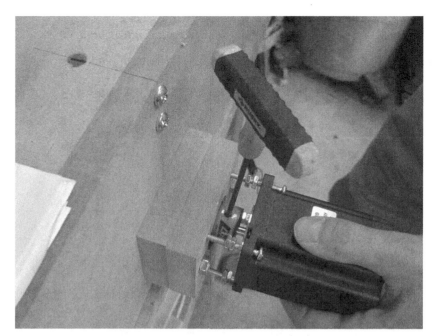

Figure 17-8. *Insert the motor shaft and tighten the couple.*

Figure 17-9. *Tighten down each nut (four to the motor and four to the motor mount).*

Examine the motor to make certain it is level and then use a wrench to tighten down the #8 nuts. Tighten them slowly and alternate frequently to keep the motor from becoming unbalanced. When done, the motor should be securely fastened to the motor mount.

Figure 17-10 shows the motor tightened down and secure on the x-axis front support (leg).

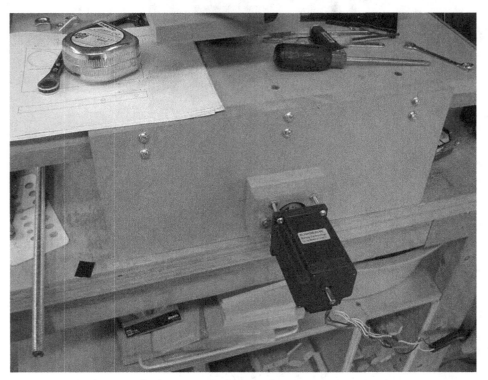

Figure 17-10. *The x-axis leg has its motor attached.*

TIP You may find adding a third nut on each #8 bolt to be useful. The third nut can be used to tighten down and hold the nut that is pressing against the motor mount (which may tend to loosen over time as the machine moves and shakes).

We know you're anxious to apply some power and get those motors spinning, but you've got one more task to accomplish before you can begin testing. You need to mount your router to the router base, Part V.

Mounting Your Router

The steps for mounting your router to Part V will obviously depend on the brand and model of router you purchase. We chose to use the Bosch Laminate Router (Colt model), also known as a *hand router.* This router comes with a thin black plastic base that is bolted to the router's metal shell—this plastic piece makes the router easier to slide along whatever material it is cutting . . . and it's a piece we don't need. So we removed it, exposing the four holes in the corners of the router (see Figure 17-11). We'll use these four holes to bolt the router to Part V.

Figure 17-11. *The router has four holes in its base that can be used to bolt it to Part V.*

Our first step is to drill the hole for the router's collet (and milling bit). We've marked the center point for drilling this hole on Part V, but this point will vary depending on the size and shape of the router you use (you can see that in Figure 17-11); Figure 17-12 shows Part V drilled with the four small holes used to bolt the router to the MDF and the larger hole for the collet and milling bit. (Again, the center point was marked based on where we wanted the final mounting of the router to be—as long as you drill your center point so that the router is not touching or interfering with any moving parts, it should be OK, but be sure to leave enough room on the router base for a vacuum hole if you wish to connect a hose for collecting dust.)

NOTE Read the instructions carefully for the router you have purchased. Each router is different, and the method used to raise and lower the collet may be different from one router to the next. We set our router's collet at the lowest level—that is, as close as possible to the material it will be cutting and drilling. You'll also want to cut or drill a vacuum hole in the router base that doesn't interfere with the router's function but allows for proper suction of the dust created.

Figure 17-12. *Part V with holes drilled to mount the router and for the collet/milling bit*

Next, we counterbored the four holes on the bottom of Part V; Figure 17-13 shows how the bolts we used for attaching the router to the base are flush with the surface (actually, just a little bit deeper).

Figure 17-13. *The router is bolted to the base.*

Finally, we bolted Part V back to the z-axis frame, as shown in Figure 17-14.

Figure 17-14. *Part V is bolted back to the z-axis frame.*

NOTE You may have to get creative when it comes to mounting your router. If you've purchased a router without holes in its base, you may cut some additional pieces of MDF to be used as a clamp to hold the router in place. These pieces would most likely be bolted to Parts W and X (one behind the router and one in front) and prevent it from moving. Be sure to check out our forum for reader suggestions and alternative methods for securing routers to the CNC machine at www.buildyourcnc.com/book.aspx.

Now take a step back and admire your new CNC machine. Figure 17-15 shows our machine, ready to go.

Figure 17-15. *The basic CNC machine, ready to go*

■**TIP** You'll definitely want to come up with some method for keeping the router's power cable and the motor wiring from getting tangled or caught up in the moving parts. One way to do this is to use a metal wire (with two large hooks on the ends—e.g., cut and bend a coat hanger) above the machine—we hooked one end to a rafter in the ceiling and used the other end to hold all the cables up and out of the way of the machine.

Summary of Work

At this point, you should have the following items completed:

Six motor mounts cut and drilled

Pairs of motor mounts bolted to CNC frame

Motor/lead screw coupling attached to lead screw (for all three axes)

Three motors bolted to the CNC frame (one per axis)

Motor/lead screw coupling attached to motor shaft (for all three motors)

Router base (Part V) drilled for collet/milling bit

Router bolted to router base (Part V)

Hardware Required

For the work performed in this chapter, you will use

#8 nuts; quantity: 12

1/4" bolts; 3" length; quantity: 6 (for attaching motor mount pairs)

1/4" nuts; quantity: 6

#8 bolts; 2 1/2" length; quantity: 12 (for bolting motors to motor mounts)

#8 nuts; quantity: 24 (for adjusting distance of motors from motor mounts)

Motor/lead screw coupling; quantity: 3 (1 per motor)

The hardware required for mounting your router to the router base will vary based on the model of router you use.

What's Next?

Congratulations! Are you ready to power up your CNC machine and test it out? Well, we'll keep this short, then—Chapter 18 is all about getting the software configured so you can test out your motors and try out a few cutting operations.

Software and Testing

Now that you have your motors attached to your CNC machine, it's time to test! We'll start out by telling you where to download the special software you'll be using to control the CNC machine. Next, we'll show you how to properly configure the software. And finally, we'll give you some simple tests you can perform to verify you've wired up everything properly and that your CNC machine is ready for bigger and better things.

CAD, CAM, and Control Software

There are three types of software that you'll be using with your CNC machine. The first is *CAD* (*computer-aided design*). This is specialized software that allows you to design two- and three-dimensional objects for the CNC machine to cut, drill, and perform other actions on. The second is *CAM* (*computer-aided manufacturing*). CAM software takes the design you created with the CAD software and converts it into a "language" called *G-Code*. This G-Code is then used by the final type of software, *Control*. Control software is the actual application that talks to your CNC machine; it takes the G-Code from the CAM software and uses it to send the proper electrical signals (via the breakout board—see Chapter 6) to the three motors.

This is a very simplified explanation of the three types of software— you'll find that in this chapter we haven't even scratched the surface when it comes to providing comprehensive details and explanations on these three types of software.

We cannot predict all the uses you might have for your CNC machine; your imagination and skills with the CAD software are really the only limit to what you can do (and your CNC machine's capabilities must be factored in). While we cannot provide detailed instructions on the use of CAD and CAM software, we are happy to tell you that the Internet is filled with discussion, photos, trial software, and more. There are numerous companies that sell CAD and CAM applications, and there are many free and/or open source alternatives. You'll probably wish to read some reviews, participate in some web discussions, and get a feeling for what software is out there that will be best suited for your purposes.

As for Control software, the same situation exists—there are free and pay-to-use software solutions when it comes to Control applications. Fortunately, we're able to provide you with a link to download a fully functional Control application that won't cost you a penny. It does have some limitations (we'll talk about those), but for the purposes of testing your new CNC machine and performing some basic tasks, you may find that the free version is all you'll ever need.

The Mach3 Control Software

The Control application we're going to be referencing in this chapter is Mach3. It's from ArtSoft USA and is available in a free version and a commercial version. Both versions are identical, but the free version is going to limit you to 500 lines of G-Code; the version you can purchase removes this limitation (although it does have an upper limit of 10,000,000 lines of G-Code).

Figure 18-1 shows a screenshot of Mach3's main control screen.

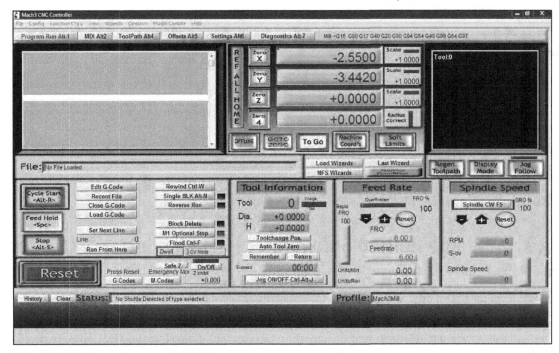

Figure 18-1. *The main screen of Mach3*

The first time you view the main screen of Mach3, you're likely to be a little intimidated; there are a lot of buttons, readouts, and other elements crowding the screen, and none of it is likely to be familiar to you. But don't worry—fortunately, there are only a handful of things you need to configure in the software to get your machine up and running. (ArtSoft USA provides a thick manual on Mach3 that you are welcome to read at your leisure—plan on setting aside a few hours or more!)

Downloading and Installing Mach3

Before we can show you how to configure the software, though, you need to download a copy and install it. So, open up your web browser, visit www.machsupport.com, click the Downloads menu, select Mach and LazyCam, and follow the instructions on the page that opens to download the installation file.

NOTE Be sure to download the lockdown version, not the development version. Also, if you're running the 32-bit version of Windows Vista, be sure to download the memoryoverride.zip file on the same page. Click the Vista README link for the extremely simple instructions on installing the patch. This will make a registry change in Vista that will allow you to run Mach3.

After downloading the installation file, double-click it and follow the instructions to install it. When you get to the screen for creating a profile (shown in Figure 18-2), just click the Next button to skip it.

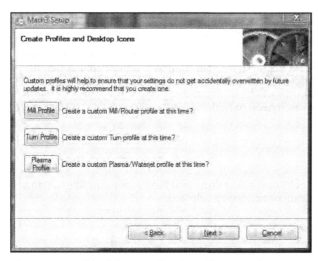

Figure 18-2. *Skip this step when installing Mach3.*

The installation will also allow you to install LazyCam, a CAM application that can work hand in hand with Mach3 once you have a CAD file that needs converting to G-Code. You can choose to install it or not; we recommend going ahead and installing it as it doesn't take much hard drive space and you may find it useful for generating G-Code.

At some point in the installation, Mach3 will want to install a parallel port driver, as shown in Figure 18-3. This is normal, so click the Next button and follow the instructions to allow Mach3 to install this driver.

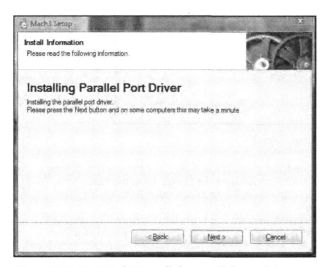

Figure 18-3. *Let Mach3 install the parallel port driver.*

When the installation is complete, restart your system and when brought back to your desktop, you'll have a handful of shortcut applications added to your desktop. The shortcut icon that we'll be using is titled Mach3Mill—go ahead and double-click that icon to open Mach3.

Configuring Mach3

We know you're anxious to take your CNC machine for a spin, so we'll get straight to the key configurations.

If you haven't opened Mach3 yet, double-click the Mach3Mill icon on your desktop. You'll see a screen similar to Figure 18-1 appear. The Reset button will be flashing, and you'll see a large collection of readouts. There aren't that many configurations you'll need to perform, and we'll walk you through each of them.

Ports and Pins

Click the Config menu in the top-left corner of Mach3 and select Ports and Pins. A window like the one in Figure 18-4 will open, with the Port Setup and Axis Selection tab selected.

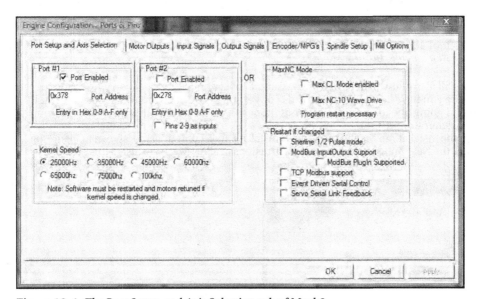

Figure 18-4. *The Port Setup and Axis Selection tab of Mach3*

The only setting you really need to be concerned about on this screen is the Port #1 box in the upper-left corner of the window. Look at the value for Port Address. In Figure 18-4, the value is 0x378—it's a hexadecimal value. You need to make certain that this value matches the port value for your parallel port.

To do this, open the Device Manager (in Windows), expand the Ports listing, right-click the LPT port, and select Properties. The port value should also be 0x378. If it isn't, change the value for Port Address to match your LPT port's value.

Motor Outputs

Next, click the Motor Outputs tab, as shown in Figure 18-5. Verify that the X Axis, Y Axis, and Z Axis rows all have a green check mark in the Enabled column.

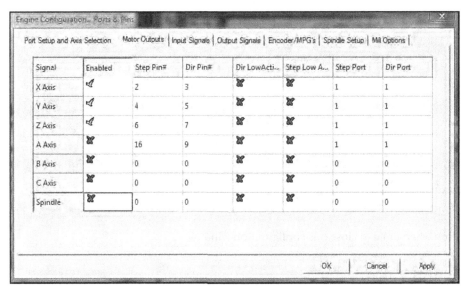

Figure 18-5. *Verify that the three axes are all enabled.*

Next, change the values under Step Pin# and Dir Pin# as follows:

For X Axis	Step Pin#: 2	Dir Pin#: 3
For Y axis	Step Pin#: 4	Dir Pin#: 5
For Z axis	Step Pin#: 6	Dir Pin#: 7

These values correspond to the terminal port numbers found on the breakout board. Remember back in Chapter 6 when you wired up all three stepper motor drivers to the breakout board in the "Wiring Motor Drivers to the Breakout Board" section. The two wires on the breakout board's terminal ports 2 and 3 connected to the stepper motor driver for the x-axis motor. The y-axis stepper motor driver connected to terminal ports #4 and #5, and the z-axis stepper motor driver connected to terminal ports #6 and #7. You use the Motor Outputs tab to specify the port numbers that Mach3 will use to communicate with a specific motor.

Input Signals

Click the Input Signals tab next. Scroll down the list until you find the EStop listing, as shown in Figure 18-6. Change the Port# value from 1 to 0. (Chapter 19 covers upgrades that you may wish to include on your machine; we'll tell you how to add an emergency EStop button that can stop the CNC machine from running when you press the button—after you've added the EStop button, you'll want to change this value back to 1 to indicate it's installed and working.)

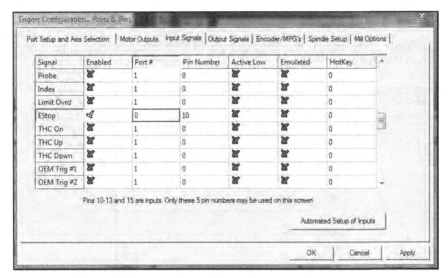

Figure 18-6. *Turn off the EStop for testing.*

Click the OK button when done to close this configuration window.

Motor Tuning and Setup

Next, click the Config menu in Mach3 and select Motor Tuning. You'll see a window like the one in Figure 18-7 open.

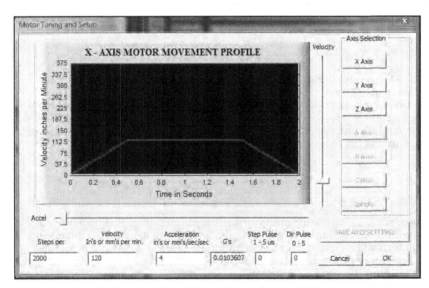

Figure 18-7. *A single screen is used to easily configure all three motor settings.*

There are three buttons of importance on this window: X Axis, Y Axis, and Z Axis. You must click a button (e.g., Z Axis) to set specific values for that motor. After setting any values, you must always click the Save Axis Settings button in the lower-right corner. (The button will remain disabled until you make a change to the x-axis motor, which is the default motor selected when you first select the Motor Tuning configuration tool.)

If you're using the motors we've recommended (see Chapter 6) as well as the microstep setting for the stepper motor drivers (also covered in Chapter 6), then you'll configure the following fields for the x-, y-, and z-axis motors as follows:

	Steps per	Velocity	Acceleration
X Axis	10400	10	7
Y Axis	10400	10	7
Z Axis	10400	10	7

Again, always be sure to click the Save Axis Settings button after making any value changes for a motor.

NOTE How did we obtain these values? The Velocity and Acceleration values can easily be changed by you, but these values were our test values that worked best for the motors we selected for our CNC machine. The "Steps per" value, however, requires a simple bit of math. You can use the following formula to calculate the "Steps per" value if you're using a lead screw with a TPI (threads per inch) value other than 13 or a microstep setting other than 1/4:

Steps per value = 200 * TPI * (1 / microstep)

So, our value was calculated using the following: 200 * 13 * 4 = 10400.

Click the OK button after you've configured the proper values for all three motors.

Configuring the Default Motor Units

This is the easiest configuration of the bunch; click the Config menu and choose Select Native Units—you'll be warned that this setting is only for configuring the motor units—click OK, select MM (millimeters) or Inches, and then click the OK button. (For our machine, we're using Inches.)

Testing Your Machine

Now it's time to test your machine! For the initial tests, we recommend you not turn on your router. There's no need to have it powered since you're not going to be cutting any material yet. You just want to test that the motors can move the router along the three axes—up/down, left/right, and backward/forward.

Connect your computer to the breakout board using a male-to-male 25-pin cable. You'll attach one end to the breakout board and the other end to your computer's parallel port.

Now plug in the power to the breakout board and the power supply. If you listen carefully, you may hear all three motors engage; it's nothing more than each motor powering up.

Open Mach3 and click the MDI (Alt-2) tab, as shown in Figure 18-8 (*MDI* stands for *manual data input*). You're going to tell various motors to move by actually typing in G-Code on this screen.

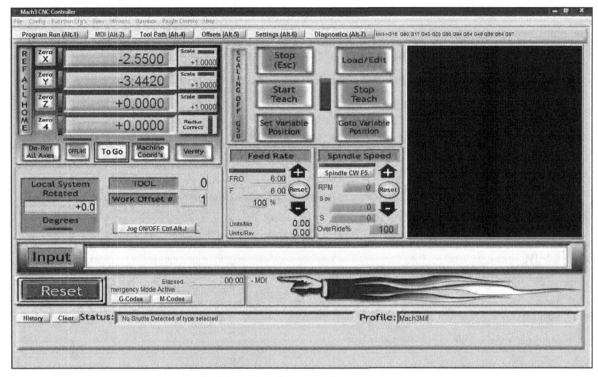

Figure 18-8. *You can test your CNC machine using the MDI tab.*

The first thing you need to do is click the Zero X, Zero Y, and Zero Z buttons to reset the initial values for your motors to zero (as pictured in Figure 18-9).

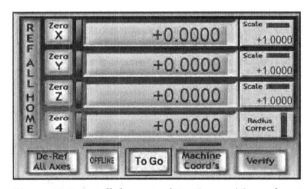

Figure 18-9. *Set all the motors' starting position values to zero before testing.*

Now, if the big Reset button is blinking (see Figure 18-10), click it; it should stop blinking. Just above the Reset button is the Input box. Click inside the text field and type in **G00 X1** (those are zeroes after the letter *G*). This is a simple bit of G-Code. When it is executed, Mach3 will instruct the x-axis motor to spin, and the router (and both the z-axis and y-axis frames) will move forward (or backward) a total of 1". Press the Enter key to execute the command.

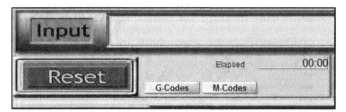

Figure 18-10. *Use the Input field to manually enter G-Code for testing.*

Run a similar command—G00 X-1—and press Enter. Did the router move in the opposite direction 1"?

Can you guess what this command is for: G00 Y1? Run it and see if you're right. It should move the router along the y-axis a total of 1". But did it move to the left or right? Make note of this, as we'll discuss direction and how to change it shortly.

For the last test, you're going to execute G00 Z1—but before you do that, make sure that your router's collet (or bit if you've inserted one) is not close to the worktable surface. If it's too close to the surface and you execute the command, the z-axis frame may move down and the router may gouge the surface. If it's too close to the surface, power down the breakout board and the power supply, and manually rotate the z-axis motor in the direction that causes the router to move up (away) from the worktable. Power up the electronics once the router is a safe distance above the worktable.

NOTE You need to power everything down because when the motors are powered, you will not be able to turn them by hand. And don't try—you can easily damage the motor.

Now execute G00 Z1 and make note of whether the router moves up or down 1". Executing G00 Z-1 should move the router in the opposite direction.

Before we test out our G-Code skills on a real piece of wood, however, let's talk about direction of movement for a moment. When you issue a G-Code command of G00 X3, do you want the router to move toward the front or the back of the table? What about the y-axis? If you execute G00 Y2, do you want the router to move to the left side of the worktable or the right side?

There is no industry standard for this type of configuration. Testing will allow you to decide what works best for you. Some prefer to configure their motor settings so that moving the router to the front of the table is a negative G-Code value (G00 X-1); some prefer to specify that moving the router bit down into a workpiece is a positive G-Code value (G00 Z4). This is one of those areas that will become clearer as you learn more about CAD and CAM software and get experience using your machine.

TIP We do have a suggestion for the z-axis. If you consider the worktable to be where z has a value of 0, then anything below the table will have a negative value and anything above the table will be positive. When thinking about drilling into a workpiece, you can use this concept to remember that moving down will be a negative G-Code value (G00 Z-1) and moving up will be a positive. G00 Z-.5 is less than -1, so the router bit will be closer to the surface of the workpiece.

Let's say that when you issue a G-Code command such as G00 Z1, the router moves down, not up. How do you change the motor settings so the z-axis motor rotates in the opposite direction and makes G00 Z1 move the router up? Easy. Within Mach3, click the Config menu and choose Ports and Pins. Click the Motor Outputs tab, as shown in Figure 18-11. To change a motor's direction, simply change the default value in the Dir LowActive column. Notice in Figure 18-11 that the z-axis motor's Dir LowActive column now has a green check box in it. This simple change will cause the z-axis motor to spin in the opposite direction. Now the G-Code command G00 Z1 will make the router move up, not down. (You can make this change on any of the motors.)

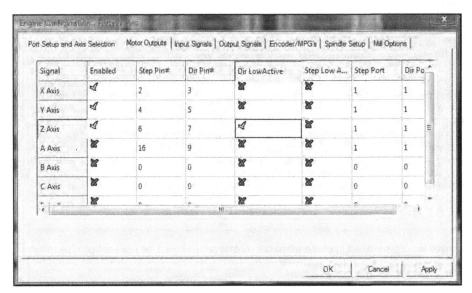

Figure 18-11. *Changing a motor's direction of spin is quick and easy.*

Testing the Router

One of the first things we did once we figured out how to control the directions of movement for the router was to map out cutting a square using manual G-Code entry. Here's the code we entered, one line at a time (after zeroing all the motors and setting the router bit's tip 1" above the piece of wood we clamped to the worktable):

G00 Z1.25: This moves the spinning router bit 1/4" into the wood to begin cutting.

G00 X2: We're going to cut a 2" square so this moves the router forward 2".

G00 Y2: This cuts a 2" line by moving the router to the right.

G00 X-2: This cuts a 2" line by moving the router backward 2".

G00 Y-2: This cuts a 2" line by moving the router to the left and finishing the square.

G00 Z-1.25: This moves the router bit up and away from the material.

Try it! Be sure to use a piece of test wood that is 1/2" thick (or greater) and big enough to allow the router bit to cut the square with plenty of spare material surrounding it. Figure 18-12 shows the results of our first test.

Figure 18-12. *Our first test allowed us to cut a square 1/4" deep into the plywood.*

School Starts

You're done . . . but only with building your CNC machine. Now comes the real fun: cutting, milling, drilling, and more. But this is going to require some more learning on your part. Our goal with this book was to give you the instructions for building your own CNC machine, and we hope we've accomplished that task and you're happy with the results.

You're going to need to learn a bit more about CAD—designing those things you wish to cut out and make using your CNC machine. CAD is one of those skills that you can spend as little or as much time developing as you desire. You'll want to investigate different CAD applications, and a good place to start is www.buildyourcnc.com/book.aspx and the forum we've set up for posting your questions, comments, pictures, and more. Internet searches are also a good way to find what you need, but be careful not to get overloaded with the sheer volume of information out there.

CAM software is yet another area you'll want to investigate; unless you enjoy entering G-Code by hand like we did earlier in the testing phase, converting your CAD designs into G-Code is just one of the functions of CAM software, and you'll want to do your research and test out various applications to find the one that works best for you.

Mach3 is free to use, but don't exceed the 500 G-Code line limit. The application is so good, however, that we encourage you to support the Mach3 application developers by purchasing the full version. You'll gain access to future upgrades of the product as well as tech support, but more importantly you'll be helping to keep Mach3 free to use for those newcomers to CNC who are just starting to experiment with their own machines.

What's Next?

To close out the book, Chapter 19 will give you a few suggestions for making your CNC machine more safe to operate as well as some references to check out for learning how to use your new machine.

Where to Go from Here

You have your brand new DIY CNC machine sitting in front of you, waiting to be put to use. If you're familiar with CNC technology, you may already be moving forward with whatever CAD and CAM software you're familiar with—but it's more likely you're scratching your head wondering what to do now.

Well, this chapter is going to provide you with some suggestions on how to get to work with your CNC machine. We're also going to provide you with some upgrades and improvements to consider for your machine.

We also recommend that you pay a visit to the book's forum at www.buildyourcnc.com/book.aspx and look to see what others are doing with their DIY CNC machines. We encourage our readers to post their questions and comments, but we're also asking for users of the CNC machine to post pictures of their projects, provide help to other users who are just getting started, and let the DIY CNC community know how they're using and modifying their machines.

Getting Familiar with CAD

Your CNC machine is made for cutting and drilling parts that you design yourself. Yes, you can easily download G-Code projects from the Internet, but the only real way you're going to become familiar with the operation of your CNC machine is to begin designing your own parts, converting them to G-Code using a CAM application, and then using Mach3 (or other Control software) for letting your machine get to work.

You can find CAD software all over the place—as many free versions exist for download on the Internet as there are commercial versions that will cost you anywhere from a small to a large fee. While the price of software isn't always an indicator of quality, software that comes with a price does usually come with technical support. With free software, you'll find that "tech support" likely comes in the form of posting questions on a discussion forum and waiting for someone to respond with a good answer.

Many of the well-known CAD applications are expensive for the traditional hobbyist—these apps often come with more features than you'll ever put to use. That's why we recommend starting out with a low-cost or free CAD application. Use it until you find it doesn't provide a feature or ability that you need; when that roadblock appears, you'll have to do some research and find a CAD application that does provide the feature(s) you need.

For a great list of CAD software, visit www.freebyte.com/cad/cad.htm, where they've put together a collection web links for free and commercial CAD applications. Plan on doing some Internet searches to obtain reviews and comments from others who have used a particular CAD application.

A few other web sites devoted to CAD that you might want to look over include

http://en.wikipedia.org/wiki/Computer-aided_design: This Wikipedia article provides a brief overview of CAD, as well as links to many more articles of interest.

www.caddprimer.com: You can download the first 70 pages of this book for free to see if it may be useful to you.

`http://avocado-cad.sourceforge.net` and `http://brlcad.org`: These are two open source software CAD applications; they're free to download and use with no restrictions.

`http://usa.autodesk.com`: This is the home of AutoCAD, considered by many to be the industry standard for CAD applications. It's expensive, but always on the cutting edge of features.

`www.cnczone.com`: Not only is this a great CNC machine web site, but it also has a discussion forum and plenty of product reviews of software and hardware.

Getting Familiar with CAM

As with CAD software, CAM is another type of CNC application that you'll need to investigate. Not only does CAM software convert your CAD designs into G-Code, but it's also responsible for generating the tool path—the controlled movement of your CNC machine's router. CAD software can reduce the amount of time your CNC machine is working by calculating the most efficient path for moving the router.

Point your web browser to `www.probotix.com/cnc_software` and scroll down the page to locate a list of CAM software for you to investigate. (You'll also find links to CAD, Control, and other CNC-related software and instructions.)

A few other web sites devoted to CAM that you might want to look over include

`www.cambam.co.uk`: CamBam is a CAM application that's free to download and use; it has a strong following among CNC hobbyists.

`www.vectric.com`: VCarve and other titles from Vectric may be a bit pricey, but have some great capabilities.

`www.bobcad.com`: BobCAD offers a free-to-try download of its popular CAM software.

Installing an Emergency Stop

We mentioned the emergency stop, often called an EStop, in Chapter 18. An EStop is a useful button for quickly shutting down your CNC machine. It's not the same as the CNC machine's power button, which may be located under the machine or otherwise not easily reachable. In the event of an emergency—such as your CNC machine hitting a knot of wood or a motor getting locked up—you want to be able to quickly shut off your machine; imagine trying to find the power button on a power strip that's under the table or out of reach, and you'll begin to understand the value of putting an EStop on your machine in an easy-to-reach location.

Installing one is fairly simple, and we'll walk you through the installation in this section. You'll want to locate and purchase the type of EStop that allows you to twist the top of the button so it releases and pops up; when the button is pressed (or quickly hit), the button stays down and won't pop back up until you twist it again to release. If you're having trouble locating one, contact `www.acksupply.com` to purchase the one shown in the following figures.

Figure 19-1 shows the items we used to create our EStop (minus the wiring).

We drilled a 1" hole in an aluminum bracket for the button and two 1/4" holes so we could bolt the bracket to the front of our CNC machine. The small black metal ring and square aluminum ring are used to secure the button to the bracket from underneath; they screw onto the button and hold it tight.

Figure 19-2 shows the bracket bolted to the front of the CNC machine.

Figure 19-1. *The components used to add an EStop to the CNC machine*

Figure 19-2. *The bracket mounted to the front of the CNC machine will hold the EStop button.*

Next, insert the button into the bracket through the square metal ring and into the hole drilled for it on the bracket. Use the round black ring to tighten down the button from underneath the bracket, as shown in Figure 19-3. Cut two pieces of wire (one black and one white) long enough to reach the breakout board. (We cut the lengths and ran them through the conduit for the x-axis motor wiring.)

Figure 19-3. *The EStop wired and ready to be connected to the breakout board*

On the bottom of the EStop, you'll see three metal posts, labeled G, NO, and NC. *G* is for *ground*; solder or connect the white wire to the G post and connect it to the Ground port on the breakout board. The *NO* label means *normal opened*, and *NC* means *normal closed*; these simply represent whether the switch is considered pressed (closed) or not pressed (open) when the power to the machine is turned on. Solder or connect the black wire to the NO port on the EStop button and connect the other end to port #10 on the breakout board. Figure 19-4 shows a closeup of the breakout board with the two wires coming from the EStop button labeled.

Figure 19-4. *The wires from the EStop button connect to the breakout board.*

Next, go back into the Mach3 software and set the EStop value to 1 (on) if it is set to 0. (Refer back to Chapter 18 for the location of the EStop configuration.) Run a test by entering some G-Code to spin one of the motors; while the motor is turning, press the EStop button, and the motor should immediately quit rotating. If it doesn't, check your wiring again for the EStop button as well as the Mach3 software setting.

Adding Limit Switches

As you become familiar with CAD software, you'll likely discover that there's no real standard for where to start a milling job. Some CAD operators start in the middle of a piece of material, with the (0,0) coordinates for the x- and y-axes as close to the center as possible. Others will set the (0,0) starting coordinates in one of the corners. You'll probably find that the starting position of your router will depend on the job you're performing.

Many CAD machines will have a set of limit switches (see Figure 19-5) to help define the starting position. Limit switches, when wired properly, send a signal to the breakout board telling the motors to stop spinning. A small arm on the limit switch, when pressed, sends an electrical signal to the breakout board, indicating the switch has been engaged.

A limit switch can be helpful for ensuring that a motor (such as the x-axis motor) doesn't move the router too far and try to go beyond the tabletop's working surface. Limit switches can also be used to define the (0,0) point on a table. For example, if you place one limit switch on the front leg of your CNC machine and a second limit switch on the left gantry, you can manually move (or use a program to move) the router to the front-left corner and define that point as (0,0) in your CAD program.

Figure 19-5. *Limit switches can be used to quickly stop the router's movement.*

You can find more information on adding and using limit switches at www.buildyourcnc.com/book.aspx, as well as search for more information in the book's forum at the same address.

Adding a Solid State Relay

Although beyond the scope of this book, adding additional safety features such as limit switches and an EStop is highly recommended. One additional safety feature that we'd like to quickly mention is called a *solid state relay* (*SSR*).

You can find more information and discussion of SSRs at www.buildyourcnc.com, but in a nutshell, an SSR is a useful device for controlling the startup and shutdown of all your CNC machine's electronics. An SSR can be used to turn on and off the router (so you don't have to turn it on manually before beginning a job), as well as disengage the motors until the program is initiated. One added benefit of the SSR is that when you turn on your machine's motors, motor drivers, and breakout board, the SSR can be used to prevent the electronics from accidentally starting up if a program is running.

Whether you use your machine often or rarely, give serious consideration to adding an SSR to your CNC machine for added safety.

Protecting and Painting Your Machine

MDF is a great material, but it's very sensitive to water. Under no circumstances should you set a cold drink on any MDF surface! Try to keep the machine in a cool, dry location and avoid putting it under or near any source of condensation or leaks. Water will cause the surface of the MDF material to swell and deform.

One good way to protect your CNC machine's MDF parts is to paint them. The downside to this is that you'll likely need to tear down your machine, paint the parts, and reassemble. Spray paint or a good coat of latex will both work—just don't put it on too thick.

What's Next?

Our goal with this book was to give you the best instructions possible for building a working CNC machine, and we hope we've accomplished that task. Now it's up to you to push your machine to its limits and see what you can do with it. You'll find the book's web site and discussion forums a good place to talk about your machine, see what others are doing with their machines, and maybe offer some advice to those just getting started with the book.

We want to see and hear from our readers, and we hope you'll take the time to register at the book's forum (`www.buildyourcnc.com/book.aspx`) and share a picture or two of your final machine—tell us about the work and any modifications you've made, and of course let us know if we've made any errors so we can fix them and inform others on the errata/corrections page found at the same web address.

Be safe, and have fun with your very own CNC machine!

Index

You Need the Companion eBook

Your purchase of this book entitles you to buy the companion PDF-version eBook for only $10. Take the weightless companion with you anywhere.

We believe this Apress title will prove so indispensable that you'll want to carry it with you everywhere, which is why we are offering the companion eBook (in PDF format) for $10 to customers who purchase this book now. Convenient and fully searchable, the PDF version of any content-rich, page-heavy Apress book makes a valuable addition to your programming library. You can easily find and copy code—or perform examples by quickly toggling between instructions and the application. Even simultaneously tackling a donut, diet soda, and complex code becomes simplified with hands-free eBooks!

Once you purchase your book, getting the $10 companion eBook is simple:

❶ Visit **www.apress.com/promo/tendollars/**.

❷ Complete a basic registration form to receive a randomly generated question about this title.

❸ Answer the question correctly in 60 seconds, and you will receive a promotional code to redeem for the $10.00 eBook.

233 Spring Street, New York, NY 10013

Offer valid through 5/10.